LOOKOUT

LOVE, SOLITUDE, and

Searching for *WILDFIRE*

in the Boreal Forest

TRINA MOYLES

RANDOM HOUSE CANADA

Library and Archives Canada Cataloguing in Publication

Title: Lookout : love, solitude, and searching for wildfire in the boreal forest /
    Trina Moyles.
Names: Moyles, Trina, author.
Identifiers: Canadiana (print) 20200276786 | Canadiana (ebook) 20200276921 |
    ISBN 9780735279919 (hardcover) | ISBN 9780735279926 (EPUB)
Subjects: LCSH: Moyles, Trina. | LCSH: Fire lookouts—Canada—Biography. |
    LCSH: Fire lookout stations—Canada. | LCSH: Wildfires—Canada. |
    LCGFT: Autobiographies.
Classification: LCC SD421.375 .M69 2021 | DDC 634.9/3—DC23

Text design: Lisa Jager
Jacket design: Lisa Jager
Jacket image credits: (front) Heather Van Haren; (back) Linda Sirko-Moyles
Image credits: (binoculars) © Pabkov / Dreamstime; tower and crosshairs courtesy
of the author

Printed and bound in Canada

10 9 8 7 6 5 4 3 2 1

Penguin
Random House
RANDOM HOUSE CANADA

Author's Note

The views and opinions expressed in this book belong solely to the author.

Some names and identifying details have been changed, but all of the stories are based on real people. Where real names are used, permission was given to the author.

All of the events that follow are true as remembered by the author, although she admits that, after one hundred and some days alone, after the seventh straight day of rain, after five fire seasons and a broken heart, the edge between fantasy and fact could often blur.

For my grandfather
John Tweed Moyles,
a lookout at heart.

For Masen and Brielle,
stay wild.

And for Holly,
who is always
down below.

CONTENTS

The lookoutman's lot is a lonely one. His place of work is often inaccessible by conventional means of travel. Most of the time the forestry radio is his only link to the outside world. Visitors are a rare occurrence. Depending on his own resourcefulness and his ability to organize his own life, it is entirely up to the individual to make his stay at the lookout a success. The man who can appreciate the advantages of a lookout life will find this work a rewarding experience as documented by the lookoutmen that have come back for many seasons and to whom the lookout has become a second home.

—*Lookoutman's Handbook*, ALBERTA FOREST SERVICE, 1971

It doesn't take much in the way of body and mind to be a lookout . . . it's mostly soul.

—NORMAN MACLEAN, *A River Runs Through It and Other Stories*

INTRODUCTION

It was early May and the last of the northwestern Canadian winter had melted away. Sap flowed through the dry, flammable limbs of the leafless deciduous trees. Resin doused the spruce and pine needles with a kerosene-like substance. The boreal forest was a box of matchsticks, waiting to be struck.

Wind howled through the bones of the fire tower, a hundred-foot steel structure that rose up between forest and sky. My vantage point was from the cupola, a tiny, red-and-white octagonal dome that measured less than three metres in diameter. I surveyed the forest below, a patchwork mosaic of aspen and birch, black and white spruce, and pine, coloured in shades of olive, ochre, marigold, ginger, and rust. Twenty, forty, eighty kilometres into the blue-and-green expanse, I became overwhelmed by the vastness of the boreal wilderness and the distinct lack of human influence. No buildings, no roads, no power lines.

My new job as a Lookout Observer, on paper, seemed straightforward enough: climb the tower every day and spend long hours sitting up in the sky, scanning the horizon for the faintest curl of smoke. Lookouts are the first line of defence in wildfire detection in Alberta. A network of 127 fire towers is scattered across the province in the boreal, foothills, and Rocky Mountain areas. From May until the end of August, I would be responsible for detecting and reporting wildfires within a forty-kilometre radius of forest. There would be no days off, no relief, no leaving the fire tower for another 120 days. And, as I'd soon learn from my seasoned neighbours, that's exactly the way lookouts wanted it. Few longed for days away from the watch. Most lookouts hoped for an early start in April and a late season extension into October. If they could, some hard-core types would probably work straight through the winter.

I had catapulted myself into a culture of vigilance that bordered on obsession. Though the backgrounds of lookouts varied—former lawyers, bankers, butchers, healthcare workers, teachers, trappers, and artists had come to the profession—there was a shared commitment to observation among them that ran deep, an expertise that was earned with the accumulation of days and seasons. Only days earlier, at my orientation, I'd felt the experienced lookouts' skepticism of me, a fledgling rookie, totally unaware of what lay ahead in her first season. I didn't blame them. I was skeptical too. The immense responsibility stretched out in the blue swell of distance. My heart flapped wildly like a hooked trout. It occurred to me that the fire tower might be one of the last places on the earth's surface where one can experience the physical reality of being entirely isolated from other people, a sensation that felt both exhilarating and terrifying. Would I be capable of enduring the summer?

To my south, a single white shoelace danced and undulated above the treetops. I focused my binoculars on the floating ribbon and saw that it was a group, or a hedge, of trumpeter swans. The white-winged migrants were flying home to their breeding ponds in the boreal forests of northwestern Alberta, or as far north as the High Arctic.

I'd come home too, to the place where I spent my childhood in the Peace River country, on Treaty 8 lands, traditional territories of the

Beaver, Dene, Cree, and Métis peoples. After years away, I'd returned to find work and support my fiancé's emigration from Uganda. The lookout job offered a stable, seasonal work opportunity. But the reasons for my return proved more complicated than love and financial security. It would take several years for me to understand what really brought me back to the North, what it means to be a lookout, and how life at the fire tower would change me irrevocably.

I grew up in northern Alberta, only a few hundred metres from the banks of the Peace River, a 1,923-kilometre river that originates in the Rocky Mountains of northern British Columbia. Wide and brown, the Peace flows northeast across Alberta to join the Athabasca River and form the Peace–Athabasca Delta. As a young girl, I played in the willows and dogwood that sprang up along the riverbanks and went searching for agates and pieces of driftwood that reminded me of the wild creatures I was fascinated by: wolves, elk, bears, moose, and caribou. My father, a wildlife biologist, studied these animals and advocated for protecting their habitats and populations against the rapid growth of farming, logging, and oil and gas exploits.

I was enamoured of my dad's work as a biologist and his adventures in the boreal forest. He and my mother, a social worker, raised my older brother, Brendan, and me close to nature: camping, canoeing, hiking, fishing, and hunting. The forest was sacred text they showed us how to read and love. My dad taught me everything he taught my older brother: how to track prey, pull the trigger, dress a grouse. We baited lines and cast and waited. He pointed out the bald eagles that plunged and skimmed the surface of the lake, talons opening and clenching, sometimes around a fish, sometimes just air. Fishing wasn't necessarily about catching fish. It was about waiting and witnessing everything else you couldn't have anticipated: a cow and moose calf emerging from the bush, a lone raven speaking an old language of throat croaks and clicks that sounded of equal parts mischief and grief.

We witnessed the wild—the nothingness and the everythingness—with all of our senses.

Every winter, when snow carpeted the earth, my dad flew in four-seater Cessna airplanes over the forest, surveying wildlife populations. On his maps, he showed me networks of red fire towers positioned throughout the boreal. I traced my finger over the towers scattered across the map, counting over a hundred of them. During the summers, my dad and his research colleagues often landed on the airstrips adjacent to fire towers to fuel the plane. He remembers being invited in for a cup of coffee by the company-starved souls who staffed the towers. Who were these people? I'd wondered. Fire towers seemed as elusive as woodland caribou, a threatened species on the Albertan and wider Canadian landscape. I'd never seen one with my own eyes. The map sparked a childhood curiosity, a fascination with what it might be like to live alone at a fire tower and watch for smoke. The map and stories stirred my imagination and appealed to my creative side. It all seemed so romantic.

"Maybe you could do that one day," my dad said. "It's a great job for a writer."

I'd always dreamt about being a writer. I was a kid who wrote and read voraciously: stories, comics, poems, essays, manifestos. In grade four, I published a hand-bound book called *When Sage Died*, an epic tragedy about the loss of my dad's loyal hunting dog, a spirited springer spaniel. Years later, in high school, I became a youth columnist for the *Peace River Record-Gazette*. I remember the tiny thrill I felt when I saw my name and words in print for the first time.

When I was fourteen years old, my dad brought home a copy of the novel *Burning Ground* by Pearl Luke, an Albertan writer who worked seven seasons at a fire tower in northern Alberta. Luke wrote her novel in the cupola overlooking the boreal muskeg. I devoured it from cover to cover, trying to imagine myself climbing a steel ladder, scanning the horizon for smoke, writing my own book, and braving the wilderness.

———

Growing up in the North, it was confusing to be a girl in what I sensed was a man's world.

In fact, I didn't want to be a girl. I ran with a wolf pack of boys, tagging along with my older brother—whom I worshipped—and his friends. We catapulted off boulders into the murky swimming hole. We piled onto the snowmobile and gunned the engine, etching our illegible longing onto the pages of snowy alfalfa fields. We built forts, climbed trees, dug around in old bear dens, slung mud at one another, and slunk home at dusk, wearing the wild proudly.

But as a young teenager, I began to understand that the world had a different set of expectations for girls. Seek shelter. Prune and preen. Keep and be kept. Home and hearth. Nurture. Tend. Be beautiful.

Prescribed feminine domesticity felt like a sharp-jawed trap. I wanted to claim the habitat of boys, ranging wide and far, and test myself against the wild. I cut off my cornsilk hair. I wore sports bras and my older brother's hand-me-down nineties grunge wear. I wanted to be brave, tough, quick-witted, and physically capable. There was a weakness I associated with being a girl, of striving to be pretty, or building dreams of weddings and babies. That girl was never me. I never wanted to be that girl.

By the summer after I graduated from high school, I was eager to leave the North. My childhood enchantment with the natural world had given way to disillusionment with small-town life and the pressure to conform. The resource economy was booming in northern Alberta in the late 1990s and a "work hard, play hard" culture prevailed, which filtered down to kids and youth. I was only fourteen years old the first time I got blackout drunk. By high school, I was knocking back straight Jack Daniel's whiskey, vodka, or any kind of bootlegged alcohol my friends and I could get our hands on. Every Friday and Saturday, we partied at gravel pits and down by the river and up atop the Twelve Foot Davis lookout. I sought validation from boys, betraying the part of myself that never wanted to be "that girl"—weak, vulnerable, desperate to belong.

Amongst my friends, there was never a question of staying in the North, a reflection of our privileged upbringings, our economic ability to leave and pursue education and careers elsewhere. We left hungry for independence and a wider world view.

I moved five hundred kilometres south to Edmonton to attend university, and years later, started a career working with non-profit organizations in international development and human rights. As I found a sense of belonging with an activist community that shared my deeper values, I began to reflect differently on being a young woman in Peace River and many of the social issues I'd witnessed: addiction, racism, violence, and toxic masculinity. I was ashamed of some of my choices to indulge in booze and boys who didn't care about me. I blamed geography for my mistakes, and labelled northern Alberta as "redneck" and backward. I swore I'd never go back to live there again. I began to fill up my passport with stamps and envision a life far from home.

I forgot about the smell of black spruce, and the vast, lonely views of the northern boreal. My dreams of being a writer took a back seat to my desire for "a successful life." That meant I'd be an executive director of an NGO, an international humanitarian worker, a global public health researcher. Even though I'd never dreamt of weddings, or babies, I supposed I'd likely be a wife and a mother. I wanted to live in the city, free from the small-town mentality I'd grown up with.

My dad never stopped encouraging me to apply to be a lookout, a job that would support my early writerly ambitions, but I now had doubts about everything I'd once romanticized. Why would a young woman, or anyone for that matter, voluntarily choose to live alone in the wilderness, watch for wildfires, and place themselves within such extreme isolation? It felt akin to the stories told by mountaineers, only there were no mountains to climb, there was only loneliness to endure— and where were the great accolades in that? Tower life seemed to offer a kind of resignation, a refuge for people who couldn't hack the real world. A last stop on the line. It might as well have been Siberia. Instead of Pearl Luke, I now thought of the writer Jack Kerouac, sitting like a monk atop Mount Desolation Lookout in the Cascade Mountains of northern Washington. Despite the popular image of Kerouac writing serenely in his notebook, the writer was tormented by the isolation. He barely lasted sixty days alone at the tower. He didn't pen the novel he thought he would, but instead scribbled madly into his notebook, cursing the bugs, the boredom, and the lack of cigarettes and booze.

A former boyfriend, who'd worked as a wildland firefighter in the Rocky Mountain House area, had told me a story about a lookout. One day, while visiting one of the fire towers in the forest, he and his crew had noticed two grave-like mounds beneath the tower. "What's that?" they asked the lookout. "I buried my soul," he said. "What's in the second mound?" someone asked, I imagine a bit nervously. "I killed a goat and buried it to keep my soul company," he said.

I'd laughed at the story, but it only reinforced my doubts about the job. "I could *never* become a lookout," I'd said, shaking my head in disbelief.

Yet—against all odds—that's exactly what I'd signed myself up for. Somehow, despite my fears and reservations, I'd found myself atop a hundred-foot fire tower, more alone than I'd been in my thirty-one years. After over a decade of rejecting the North, I'd finally come back to the land that had shaped me. The truth is, I'd begun to crave the wildness of my childhood, to surrender once again to a homing instinct. And then there was love. A desire to start a new life with my fiancé and eventually bring him to Canada too.

But the fire tower didn't feel like home—not yet, anyway. I couldn't recognize in myself the girl I used to be, nor the woman who I'd become. I was afraid of facing myself, the past decade of my life, and the recent choices that had led me back to the Peace Country, far away from my fiancé, Akello, and the life we'd built together. It felt as though I'd been dropped off on another planet and traded in my former career, relationships, and life for one sole task: watching for smoke from wildfires burning out of control.

———

Ground fires in the boreal forest are dangerous because they burn unseen, insulated by layers of heavy peat and muskeg. These surreptitious fires can sleep and smoulder for days and weeks, hidden beneath the earth. But under the right conditions, on hot, gusty days, ground fires can come to life, awakened by the wind, shaken free from their roots.

For the past ten thousand years, the boreal has followed an ancient cycle of seasonal extremes. From November to April, the spongy boreal muskeg soils freeze solid, layered by snow and ice, and the flow of sap in the trees slows to a near halt. For months, the breath of the forest is barely audible until spring arrives, unlocking the blood and pulse of the boreal. By May, light swarms the northern skies and heat lures life back to the North, and the forests bake and dry and ready themselves to do as they've done for thousands of years: burn. The tightly bound coniferous seeds, cones poised at the tops of trees or the tips of branches, are desperate for release and regeneration. The forest succumbs to sudden acts of Mother Nature—soaring temperatures and bolts of molten hot lightning. Or to the clumsy hands of humans, who unknowingly, or accidentally, cause the majority of wildfires in the boreal forest. One strike. One mistake. That's all it takes for ignition, for wildfires to let rip across these lonely landscapes. The aftermath of wildfire in the boreal can easily be perceived as destructive, but it's a natural phenomenon. Fire can also be ecologically regenerative. The spruce- and pine- and bark-dwelling insects and birds and mammals have been designed by nature to thrive and rise from the ashes. The boreal is an ecosystem shaped by wildfire. The old forest must burn in order for succession to take place.

Historically, wildfire cleansed the forest of dead, diseased, and dying trees. First Nations peoples in northern Alberta traditionally practised "cultural burning" on the landscape, using fire as a tool to transform old forest into grasslands and meadows to create new habitat for deer, moose, and elk, control rodent populations, and stimulate berry production. They'd set fire to the land in early spring or late fall, when the conditions were damp and wet and fire could smoulder slowly—at lower intensities—so as not to burn too hot and damage important cultural plants. But Indigenous burning practices were outlawed in Alberta, even criminalized, in the early twentieth century. Fire as an "evil" entity began to seep into the settler consciousness, particularly following WWII, when forestry operations became militarized. A century's worth of forest mismanagement in Canada—the agenda of supressing wildfires, even small ones—has created a vast expanse of old forest that is literally dying to burn.

According to top climate scientists, the stakes for wildfires are changing drastically. The temperatures are rising and drought is intensifying. The forest is older and drier. In the northern boreal, the fire season has lengthened by more than twenty days a year. There are also more people on the Canadian landscape than ever before, which creates a higher probability of both people starting fires and people being in the way of wildfires. Scientists today agree: wildfires are burning hotter, faster, and more furiously through the forests, scorching not only what's above ground but also potentially deep into the earth, burning trees, plant vegetation, roots, muskeg, and wildlife that can't escape, right down to the mineral soil.

I'd come to face my truths, smouldering like a ground fire, at the tower. I'd see the smoke on the horizon, sense the wildfire burning out of control. I'd learn how everything you think you know about yourself can go up in flames, and be burned into ashes, so quickly. And yet I'd also witness the way fire powerfully transforms the boreal landscape and catalyzes regeneration.

On a hot, windy day, the newborn flame stirs, licks, sucks, crackles on the low-lying brush and skeletal branches. Trees torch like monks in protest. The wind shakes embers from their robes and the fire becomes unstoppable, carving out a hungry, haphazard trajectory. Within seconds the smoke can rise so black, voluminous, and terrifyingly animal, you can't look away. You can't see anything else. There is nothing for you to do but witness the smoke gathering, growing, accumulating into massive columns, staining the sky. This is how wildfires are born and witnessed in the boreal forest. This is how nature makes way for regeneration—life rushing, rushing, rising from the ashes.

PART ONE

IGNITION

[Some fires] light spontaneously, or start when lightning stabs deep into the ground to ignite layers of dead leaves with one quick strike. Yet others begin when lightning hits the heart of a tree determined to burn from core to roots like an oversized wick.

—PEARL LUKE, *Burning Ground*

CHAPTER ONE

In January 2016, with some trepidation, I submitted my application to
the Government of Alberta for the position of Lookout Observer.
Before hitting Send, I'd read a bare-bones job description:

> *A Lookout Observer's first priority is to provide early detection and
> accurate reporting of potential forest fires. A successful candidate
> will be in excellent physical and mental health to withstand the
> rigors of climbing the tower and generally living alone. Only
> highly self-motivated individuals can overcome the loneliness and
> often monotonous routines of the lookout observers' way of life.*

Going to work at a fire tower—committing to loneliness and monot-
ony—had once represented going backwards to me, a kind of failure.
But now, as I clicked Send on the application, that disdain was gone,
in its place anticipation mixed with fear. It had taken me a long time
to get here.

After leaving Peace River for Edmonton when I was eighteen, I began to travel even farther south with a hunger to see the world beyond my childhood in the northern bush. This desire, in part, was inspired by my mother's work in community rehabilitation. For decades she worked tirelessly with non-profit organizations that provided access to early childhood development services and education in the Peace Country. She often travelled to small First Nations communities to teach social work courses at remote college campuses. I admired my mother's selfless dedication to front-line work with rural communities, and her broader lens for understanding social injustice.

When I was twenty-two years old, I spent a year in northern Nicaragua, working with a mural arts organization, which engaged children and youth in transforming the blank walls of their city into works of public art. FUNARTE came together in the 1980s, following a decade-long civil war in Nicaragua, helping young people to think critically and creatively about their environments, to express their thoughts and desires. My work in Nicaragua further embedded in me a fierce sense of idealism with respect to social and political change. I wanted to make a difference in the world, to contribute to something larger than myself. The following year I came back to Edmonton and landed a job with Change for Children Association, a social justice organization that supported rural development and human rights projects in Latin America and global education projects in Canada—a dream job for my twenty-three-year-old self. I connected to a vibrant community of other social justice organizers, who would become some of my closest friends. My supervisors were three inspiring women who had committed their lives to the struggle for human rights, locally and globally. One day I'd be organizing workshops where youth could focus on the human rights issues they cared about and produce their own documentary film projects, or radio shows, or magazines; and the next I'd be speaking to a roomful of philanthropists about why they should donate to support organizations like FUNARTE in Nicaragua.

I worked with Change for Children for nearly six years, leading groups of university students into remote rainforest areas in Nicaragua and Guatemala to help communities build schools, libraries, and health clinics. I helped translate and facilitate cross-cultural exchanges and sustainable agriculture projects with Canadian farmers and farmers in Cuba.

These early years were dedicated to rethinking the way we design our communities, to joining grassroots efforts of solidarity to help communities work their way out of poverty and achieve basic human rights: the right to clean drinking water, education, food security, and meaningful, safe, equitable work.

I learned Spanish and studied cultural sensitivity, translation, conflict management, logistics planning, and project management. Many other young people reached out to ask me: How can we do what you do? How can we work in human rights? I knew I was lucky to have landed such a job and lifestyle. But there were particular challenges that came along with working in the non-profit sector. Namely, I was burning out.

The pay was minimal, the hours long and exhausting. The idea of taking time off work for self-care seemed selfish. I didn't realize that I needed emotional, even psychological support to help myself debrief and make sense of disparities I encountered through my work in Central America, meeting with people who'd lived through unthinkable poverty and violence. I didn't know how to say no to more projects and responsibilities. In fact, I added more work to my plate through my volunteerism. In 2010, before graduating from university, I co-founded a youth justice organization in Edmonton called CEIBA, taking on major board and organizational responsibilities. In my mid-twenties I began to adopt a righteous, even martyr mindset about my work. I thought of my *compañeras*, colleagues, in Latin America who were struggling against greater challenges than I was in Canada. It fuelled my work. *Patria, o muerte!* Country, or death!

But amidst the frenetic work pace, and despite my determination to carve out a space away from Peace River, the North still exerted a pull on my consciousness.

————

A woman went missing from a fire tower outside the town of Hinton, Alberta, in the summer of 2006. The story made the national news. I was twenty-one years old, travelling with a team of university students to a small village in eastern Guatemala, helping to haul buckets of water from a nearby stream and mix concrete to build a medical clinic. I was thousands of kilometres away from home when I learned about Stephanie Stewart from my father.

When he told me about the disappearance over the phone, a chill washed over me. I fumbled for the right words, not sure what to say.

"She was seventy years old," my dad continued. "They haven't found the body."

Stephanie Stewart disappeared from her fire tower on August 26, a few weeks before the end of the fire season. When she didn't answer the radio, Forestry sent out a ranger to check in on her, but she was nowhere to be found. A large pot of water had been left on to boil, possibly the water she'd planned on using for bathing. Items missing from her cabin, the RCMP discovered, included a red-and-blue Navajo-patterned blanket and an expensive watch that had senti-mental value for Stephanie. She had been a lookout for eighteen years, and was an adventurous woman who'd climbed Mount Kilimanjaro and ridden her bicycle across Canada. According to her closest friends, Stephanie loved to read and paint and do needlepoint. She'd even worked a summer up at one of the remote fly-in towers in the Peace Country. Later, I'd hear a story about how, during that fire season, she bravely took down a grizzly bear that was trying to break into her cabin. *BOOM!* A single shot through the screen door. After her disap-pearance, her daughter was quoted in a newspaper article as saying, "Mom's a hell of a woman. She's very strong, she's very capable. The tower life is her life."

Stephanie was a woman who had never needed rescuing.

The RCMP and forestry representatives from the Government of Alberta exhaustively searched the area but couldn't find any sign of Stephanie, or evidence of what had happened to her. The tower where she'd gone missing was surrounded by mountains and tightly packed forest cover. The terrain was rough and difficult to access.

The unsolved mystery of Stephanie's disappearance always stayed with me, as have the numerous stories about the girls and women who've been reported missing in northern Alberta and British Columbia. I grew up driving the same winding country road in Peace River where a family friend's younger sister disappeared in 1984. Carolyn Pruyser was only eighteen years old. A university student home for the summer. Her abandoned car, with the keys dangling from the ignition, and her purse, were all the police recovered. Although she disappeared the year before I was born, her story haunted me as a child, and more prevalently as a teenager.

The Highway of Tears, a 725-kilometre corridor of Highway 16 between Prince George and Prince Rupert in northern B.C., is the site of many murders and disappearances of girls and women—over half of them Indigenous—that have taken place since at least 1969. The bodies of these women have rarely been recovered, their cases mostly unsolved. While gender-based violence transcends race, there is a disproportionate impact on Indigenous, Black, and Brown women in Canada. In recent years, First Nations' and Aboriginal groups successfully lobbied the Government of Canada to launch a national inquiry into the staggering numbers of missing and murdered Indigenous women. Indigenous women are 12 times more likely to be murdered, and 16 times more likely to disappear, than any other demographic group in Canada.

The story about Stephanie's disappearance at the lookout—and the broader context of gender-based violence against all women, especially in the North—reaffirmed to me what I'd always sensed about my body from a young age: it wasn't always safe to be a girl or woman. Not in the North. Not at a fire tower. Not anywhere in the world.

In 2010, when I was twenty-five years old, I'd been hired by an Edmonton organization to lead a library construction project with a team of Canadian university students in northern Nicaragua. When the project wrapped up, my colleague, Michelle, and I spent some time visiting friends in Estelí, the same city where I'd lived and volunteered with the mural painting organization. One morning we boarded a bus heading back to Managua in preparation for our flights back to Edmonton the following day. We were both supposed to don blue gowns and hats and walk across the stage for our university graduation several days afterwards. On the bus to Managua, we befriended a young man, our age, who claimed to know some of my friends in Estelí. After chatting for nearly two hours, we accepted his offer of a ride to our hostel. Taking taxis in Managua was risky—better to accept a ride with a new friend. I'd made many friends while travelling through Central America in this same way. Strangers became friends, and friends helped one another out.

As Michelle and I stood to follow our new friend off the bus, I caught a glimpse of a tattoo on the man's bicep. For a split second, I hesitated. Was it a gang tattoo? These are the moments when we catch sight of the future, sense danger—feel the prick of intuition—but I ignored my gut instinct and got into the back seat of his brother's car.

Moments later, three people—a skinny man with a pink bandana around his head, a stylish woman with large hoop earrings, and another woman who looked eight months pregnant—rushed into the vehicle. All three of these people had been sitting strategically around us on the bus. I was pushed up onto the laps of the strangers as the driver took off. Just as I wondered what was happening, the man with the bandana turned around, pointing a switchblade at me, gesturing with the knife.

"*Quita la ropa*," he said.

Those three words sent me reeling into disbelief.

Take. Off. Your. Clothes.

"W-w-what's happening?" asked Michelle, who didn't speak Spanish.

"They're going to rape us," I said calmly, unexpectedly. A shot of clarity: if they raped us, maybe they wouldn't kill us.

I could live with being raped, I rationalized, coldly, with myself. But I wasn't ready to die. The thought of my parents discovering the news, realizing I wasn't on my flight back home, dealing with the Nicaraguan authorities to search for my body, was completely overwhelming. I felt more animal than human. My neurons fired wildly. *They are going to kill us*, I thought, over and over again.

"No, no, no, no," I cried, pushing away their prying fingers that were trying to take off our clothes, insisting we had no money in our pockets. No moneybelts. "Everything you want—it's in our bags."

What did they want? Money? Cameras? Our bodies? I prayed our clothes would stay on. Not knowing was torturous. I heard myself whimpering in the back seat of the car as the man who'd tricked us forced a hand over my eyes, stroked my hair, and whispered, "Shhhhhhhhhh."

"*Ella es linda*, she's pretty," the pregnant woman crowed. Michelle was wedged in between our captors in the back seat and my own body was laid out overtop of everyone's laps. The flimsy yellow cotton blouse I'd put on that morning rode up on my stomach and I was exposed, on display. Someone wrote on my stomach with a ballpoint pen. The pen struggled across my flesh. I heard a camera shutter go off. It was Michelle's camera. They all laughed. Later, I'd read what they'd written, trace the word with my finger.

Z-O-R-R-A. Slut.

They wanted money and demanded the PIN numbers of our debit and credit cards. Michelle had the project's credit cards, so between the two of us we had six cards. That was six sets of PINs to translate to our captors. When Michelle stammered, uncertain, suddenly, of the project card's PIN numbers, they pummelled me—the translator—with their fists. Someone slammed a fist into my thorax and all breath exited my lungs. The pain was stunning, then piercing.

They held us for six hours. We drove through some of the busiest neighbourhoods in Managua, I suspected, stopping at ATM after ATM, until they withdrew to every card's maximum limit. And then I could tell we were leaving the city, as the noises of people and traffic, and the tight turns, faded then diminished. Eventually, the tires crunched over gravel and the car came to a halt.

"Get out!" screamed the woman in the front seat. "Don't look or we'll fucking shoot!"

Someone shoved a gasoline-soaked rag over my mouth. Then they pushed me out of the car. I groped for air, then grabbed on to Michelle's flailing limbs. We held on to one another, crying. This is it, I thought. This is the end.

The driver gunned the engine, tires spitting rocks, and our captors sped away, leaving us at the side of a gravel road that ran along the edge of a cornfield just off the main highway. Stunned, we staggered back to the highway and asked an old man, a farmer, "Which way to the city?" He didn't say a word, but pointed left with his chin. We began to walk along the narrow highway shoulder back to Managua, completely disoriented as traffic rushed by. A passenger bus, an old yellow school bus with the words *VIAJO CON DIOS*, "I travel with God," painted on the side, screamed by us, coming dangerously close.

I knew we had to stop and try to call for help. We stumbled into a fish-packing plant and asked to use a phone. I'm sure that the workers knew, just by looking at our dazed expressions, what had happened. An hour later a truck pulled up and our Nicaraguan-Canadian friend, Mario, jumped out of the driver's seat and came running. As we flew into his arms, I finally allowed myself to cry. I could feel my knees threatening to collapse. It was over, I thought. The longest six hours of my life.

But it wasn't really over. It was only the beginning.

Back in Edmonton, in the months following the kidnapping, I would be diagnosed with post-traumatic stress disorder (PTSD). Living in the aftermath of assault is not really living; it's waiting for something bad to happen. It's the feeling of being perpetually afraid. My brain had become more reptile than human, transformed by trauma, the shock you feel before you think you're going to die. I didn't realize it at the time, but my PTSD brain had become hard-wired to crave adrenalin: high emotions, high drama, high stakes. I struggled to hold down my job with a non-profit organization. My relationships—romantic and platonic—began to blow up in my face, one by one, often ending in an alcohol-soaked mess. I was addicted to stress, even though stress left me, physically, emotionally, and mentally, torn to shreds.

Two years after the assault, in the summer of 2012, a medical doctor, Dr. Adroa, whom I'd met through my work with non-profit organizations in Edmonton, offered me a position with his health care organization in Kabale, a small town in southwestern Uganda. I had been feeling disconnected in Edmonton, and maybe even depressed. Although I was seeing a therapist, the effects of the assault and my PTSD were always lurking just below the surface. It often felt as though the assault was the most significant thing that had ever happened to me; on bad days, I worried that it defined me. I'd also become restless in my job and wasn't sure what to do next, so I accepted the job offer, dropped everything, and booked a ticket. Time to go.

I flew to the city of Entebbe in central Uganda via London on New Year's Eve.

What I remember after landing in Entebbe: the humid black night falling heavily on my skin, the prick of stars through a dense cover of smoke from a million charcoal cooking pots. My own disbelief. I'd never felt more out of my geographical element, but I felt at home in the familiar anticipation of adventure, of not knowing what was going to happen. Dr. Adroa had sent a representative from his organization, a social worker named Lilian, to pick me up at the airport. Lilian sat up front with the driver, and I sat in the back seat as we drove along Lake Victoria towards Kampala, the capital city, passing people on the side of the road—men pushing bicycles loaded with mammoth bunches of green plantains, women with kaleidoscope colours wrapped around their waists, carrying baskets of food on their heads as they walked home from a long day at the market. As we drew into the mouth of the city, the traffic of sedans and micro-vans thickened. Boda bodas, or motorcycle taxis, darted in and out of the jams, their female passengers sitting sidesaddle, no hands, eyes glued to their cellphones, feet dangling.

I held on for dear life.

Lilian dropped me off at a fancy hotel in downtown Kampala. The hotel was virtually empty. I could hardly afford to stay here, but it was 9 p.m., I was delirious with jet lag, and I didn't want to disappoint Lilian, who had gone out of her way to book the room for me. I handed over a crisp American hundred-dollar bill to the hotel front desk and rode an elevator up to the fourteenth floor. By 4 a.m., voices speaking in Luganda, Swahili, Runyankole, and English roused me from sleep. I peered out the window and looked down. I had never seen so many people in one place before. Everywhere I looked, bodies swarmed. It was as if someone had poked at an anthill.

Uganda is one of the most densely populated places on the planet. A country that could fit into Alberta nearly three times, its population density is surpassed only by India and China. Here, space was a luxury. Here, space was a one-hundred-dollar hotel room and a view from high up. Take a good look, I thought. It was the last bird's-eye view I'd have during my three years in Uganda.

The following day, on a dilapidated charter bus that was packed to the brim with passengers, bags, and boxes of vegetables, I rode elbow to elbow from Kampala to Kabale for eight hours. I was wedged between Lilian and a silver-haired pastor named Caleb, who stood up before we departed and announced, "Let us pray for the journey." The bus journey was a blur of bodies and conversation and colour as the land rolled by, the poorly paved highway disintegrating, ground into dust the colour of chili powder. I felt drunk on greenery: the tattered banana leaves, the tidy hedges of tea plantations, the sharpened tips of king grass, the swaying sugar cane, the maize leaves fluttering like party streamers.

We stopped after several hours of heaving along the highway, and I followed Lilian off the bus. She bought a bag of sliced pineapple from a roadside vendor to share. I was already learning my name. "*Omuzungu!*" ("Foreigner!") cried a small boy wearing sandals made out of an old tire.

It was January and my skin hadn't experienced sun and heat for four long months. I was as white as a maggot beneath a log, exposed, out of place. I looked up into the branches of a large tree and gaped at the sad, slumped figure of an old man with a bald head wearing a black suit jacket. Only it wasn't a man, it was a bird. The marabou stork, one of the largest land birds on the planet. *Marabou* derived from the Arabic word *murābit*, meaning hermit.

The streets swarmed with people, but I couldn't take my awestruck eyes off the stork. What are you doing here? I wanted to ask the old man hermit. He leered down at me with a long, pointed beak. His eyes were black, old, and knowing. What are *you* doing here? he seemed to say.

What *was* I doing?

I'd stay for three months, I told myself. I'd volunteer with the organization, and wait to hear about my grad school application to an agro-ecology programme in Norway. I'd never worked in East Africa before, and I was eager to learn from Dr. Adroa and his organization.

Maybe only the stork knew the truth, that my longing was so old and so deep that I had flung myself across ocean and continent. That I would do nearly anything to lose myself in unfamiliarity. I had come to Uganda to dissolve my sense of self into the throngs of people, into the new landscape. I wanted to run away from my story of PTSD and depression, to forge a new path.

After settling in the town of Kabale, I poured my heart into my communications work with a local non-profit organization that ran a primary health care and HIV/AIDS outreach programme in the surrounding villages. The organization was just starting to develop a branch focused on malnutrition and food security on the home and community level, which they invited me to help build from the ground up. I applied for funding grants, researched food security solutions—including rabbit breeding as a low-cost, low-risk approach to helping families consume more protein—and learned about the local issues that families, many of them subsistence farmers, were facing. I began my career as a freelance journalist and writer, and published my first article, focused on child and maternal health, with a publication called *Think Africa Press*. It reminded me of the thrill I felt when I was an adolescent, writing for the local paper in Peace River. I started to write regularly for travel and human rights journals and magazines. And eventually I decided I wanted to write a book inspired by the women farmers I'd met through my work in southwestern Uganda.

Three months turned into six months, and after nine months I met Akello, whose warmth, kindness, and commitment to human rights truly inspired me. Our friendship developed slowly over my first year in Uganda, and eventually we fell in love. After another year of living together, we decided to get married. *Ma pati*, I called him. In his mother tongue, Lugbara, it meant "my tree." He grounded me. He weathered my moods. He became the root I'd lost in Nicaragua, the part of me I didn't know how to get back.

We drew up plans to build our home on a slope covered with pineapple plants. We'd already picked out the names of our unborn children: *Ayikosi*, Happiness. *Alesi*, From Love. Akello and I envisioned a life together in both Uganda and Canada. We wanted our children to know Uganda, but we also couldn't deny the opportunities in Canada—the access to education and public health care, the ability to find good employment. We'd live permanently in Canada—or so we hoped.

After flying back to Edmonton in 2015, I made an appointment with an immigration lawyer to begin the outland spousal sponsorship process.

"It's going to be difficult to collect all of the necessary paperwork from Uganda," she said matter-of-factly. "I helped one couple from there. It's possible. But be forewarned, it's going to be a nightmare."

Her words confirmed what I already knew to be true. The year before, Akello had applied for his driver's licence in Uganda. He paid the fees, took the exam, and passed, but it took the authorities four months to give him his licence. When it was finally issued, he took the bus, travelling three hours north to Mbarara, to pick up it up. Twice, the secretary at the registry office told him, "The boss isn't in, come back tomorrow."

"He's just fishing for tea money," Akello fumed. *Tea money* being a Ugandan euphemism for a bribe.

"Well, just give it to him, then," I urged.

After bribing the official, they gave Akello his licence and he took the bus home, only to realize that the licence he'd been issued had the wrong year of birth. The lawyer explained that the Canadian spousal sponsorship application would require every kind of documentation imaginable: blood work, medical clearance, pay stubs, passport, bank account information, employment records, letters of character, and police record checks to determine that he was not a "terrorist." I would have to construct a scrapbook documenting every single detail of our relationship to provide evidence of our courtship, how long we'd been together, when we moved in together, and the marriage proposal, together with photographs of our engagement rings.

"Nothing is ever guaranteed," the lawyer warned me. "You could wait five years to hear back. They could deny your application and you'd have to start the process all over again."

Five years.

As the sponsor, I would have to prove my residency—and job security—in Canada. Akello wouldn't be allowed entry into the country, not even as a tourist, until approval of our application.

I considered my only alternative—one that Akello and I had put enormous thought into. We could stay in Uganda, make a life there.

And what a good life it was. I adored our home and community in Kabale. We had a vibrant network of supports: friends, family, and colleagues. If we had children together, we'd have plenty of help with child care. The more flexible work culture would allow for time to dedicate to my writing. The cost of living in Uganda was significantly less than in Canada, which would alleviate financial pressures.

But there were more sobering facts about Uganda, including the reality of widespread unemployment amongst young professionals. I had enjoyed my former job as a liaison between the Ugandan medical professionals and the American public health students, but I'd already trained one of my Ugandan colleagues to take over the contract. I wasn't sure if I could find other well-paying NGO work in Kabale. Similarly, it would be a challenge for Akello to find a decent-paying job as a mechanic. He could try to open his own garage, but he lacked the hard skills and experience necessary to do so. Entrepreneurship wouldn't happen overnight.

What scared me the most about staying permanently in Uganda, though, was the lack of access to quality health care. Despite meeting many brilliant and dedicated Ugandan doctors and nurses, I recognized how the public system was—compared to a Western standard—in partial disarray. I'd seen the number of labouring women lined up outside the Kabale Hospital's maternity ward. There weren't enough beds, nor enough physicians, to care for the number of women in need. Women often brought their own mattresses and endured early labour outside in the courtyard, attended to by female relatives. Akello and I knew we wanted a family, and I had to consider the risks of giving birth and raising children in Uganda. It had one of the highest maternal mortality rates in the world: 343 women per 100,000 live births die from pregnancy-related causes, whereas, in Canada, there are only 8 deaths per 100,000 live births.

The truth was that, even with access to private health care, Uganda's health care system wasn't on a par with Canada's, where good-quality care is mostly free, and far more accessible.

For practical reasons, for the health of our future family, I wanted the option of returning to Canada. And though I loved our life in Kabale, it was hard to imagine a permanent estrangement from my home country. I wanted Akello to know it, too. I couldn't wait to introduce him to the Canadian winter, go snowshoeing and cross-country skiing, drive north to Peace River and lead him in the footsteps of my childhood in the boreal. If it took us five years, or longer, to get there, so be it. I began to prepare the steps to support Akello's immigration to my home country.

CHAPTER TWO

In late January 2016, a few weeks after submitting my application, I was invited to interview for a fire tower position in Peace River. Days later, I found myself sitting across from two wildfire rangers at a boardroom table in a small meeting room in the forestry office.

"Let's say two hunters show up at your tower and refuse to leave," said the male ranger. "What do you do?"

The hairs on the back of my neck pricked up, as though someone had opened a door and let the subarctic January air filter in. My mouth went dry. I rummaged around for the appropriate response, but all I could think was: Really? This is the first question? Not, "What are your strengths and weaknesses?" or, "Tell us about a challenging experience that you overcame," or even "Why do you want to do this job?"

The very thought of being alone in the bush and dealing with two armed men touched at my most real fear. I froze like a whitetail deer caught in headlights, my mind catapulting back six years to the kidnapping in Nicaragua.

As I mentally replayed those events, the ranger was studying me, waiting for a response. First question and they'd already caught me: an imposter. Women like me didn't belong alone in the woods. What was I doing here? I stared back at the ranger and scrambled to land on a logical answer, but in truth I wasn't sure what to tell him. I hadn't even seen a photograph of a fire tower before. I didn't know what tools were on site to resolve such a situation: a phone, radio, a shotgun?

I said something like, "I'd try to put space between myself and the men. I'd climb the tower. Lock the cabin door. Call for help."

What I wanted to say but didn't: *Poke the muzzle of my shotgun through the screen in the window. Start counting down from five.*

The interview would go on for two long hours. It was probably the longest interview I'd ever sat through, and definitely the strangest. The more questions they asked, the more doubtful I became. What was I getting myself into?

"You have a disagreement with your tower neighbour. How do you solve the conflict without ever coming face to face with them?"

"You cut yourself on the lawn mower blade. What do you do?"

My first thought: I have to mow a *lawn* in the middle of the woods?

"What do you do if your refrigerator breaks down?" asked the female ranger.

That one was easy. For the three years I lived in Uganda, I never had a refrigerator. What would be the point? The power went out several times a day, on average. Power outages could last anywhere from thirty minutes to twelve hours. Akello and I chilled yogurt and juice in a basin of cold water and stored root vegetables on the cool concrete floor. Eggs were fine at room temperature. Daily, we bought milk and boiled it on the stovetop. We were used to relying on rice and beans.

"Your supervisor drops off a bush-hog at your tower. What do you do?"

I had no idea what a bush-hog was, but I assumed it wasn't some food security scheme on the part of the government. It was probably a piece of machinery. "I guess I'd ask him what it does and how to use it," I shrugged.

The ranger scribbled something on his page.

"Do you plan on foraging for wild foods at the tower?" asked the ranger, and they peered at me a bit suspiciously, perhaps trying to gauge if I was one of those urbanite back-to-the-land hippie types. Another ranger would later tell me that they try to "weed out" any applicants looking to re-create a David Thoreau, or Jack London, or Chris McCandless experience in the wild. McCandless, whose life had been documented in the book *Into the Wild* by Jon Krakauer, had tried to survive alone for three months in Alaska. After subsisting on a diet of wild plants, berries and squirrels for 113 days, he died of starvation. Although it's not entirely conclusive, there is speculation that Chris accidentally ingested toxic seeds of the Eskimo potato plant, which interfered with his body's ability to metabolize food. I'd watched the film adaptation, horrified by the scene of a gaunt-cheeked Chris zipping himself into his sleeping bag, lying back on the mattress in an old bus converted into a trapper's cabin, and the life vanishing from his wide-open eyes. The government didn't want any similar types of tragedies on its hands.

"No," I said, because it seemed like the answer they wanted to hear, even though the truth was that my hands longed to caress a meadow of wildflowers and herbs, or to pluck wild, jewelled raspberries from heavy boughs. My eyes thirsted for the abundance of the forest where I spent my childhood, my tongue to name, identify, and give language to what's wild.

And finally, they asked the question I thought they'd ask from the very beginning. "Why do you want to become a fire tower lookout?"

My motivation was less romantic than my younger self would've imagined. Simply put, I needed the money. I wanted to bring my fiancé to Canada. I needed to go back to Uganda in the fall when the fire season was over. Also, I wanted to know if I could do it—if I was strong enough to endure the psychological challenge of being alone for four months. I didn't tell them about that last part, that curiosity, or unknowing. I didn't want to give them a reason to doubt that I'd survive a season.

And although I couldn't admit it yet, I was tired. I wanted a break from the burden of my responsibilities and the uncertainty Akello and I were facing together. I wanted to forget about bureaucracy and borders. I wanted the chance to catch my breath. To write from the sky. I was already doubting my choices, feeling the pressure of a relationship that demanded so much from me, and bearing the responsibility for our shared future. I wanted to remember who I had been before so much happened in my life. I wanted to feel alive again.

The hardest part about reintegrating into Canada after three years living in Uganda was the space. Not the space that lived beyond cities and towns, the rolling grasslands and northern spruce and the big, empty skies of a thousand colours. What I couldn't handle was the emotional distance—real or imagined—that seemed to resurrect continents between both strangers and loved ones.

For the first year after my return in early 2015, before I moved back to Peace River, I lived and worked in Edmonton. When I walked down Jasper Avenue in downtown Edmonton, I felt lost in a cold, colourless dream where people skimmed by one another, without touching, or exchanging eye contact, or saying hello.

I'd missed significant events in my friends' lives: graduations, new jobs, engagements, weddings, and babies. And I'd changed, too. We were all older, and everyone was busy negotiating their complex and uncertain worlds. I struggled to connect with people I loved, even when we were sharing tea or coffee, sitting at the same table, instead of an ocean's distance away. My older brother lived in Edmonton, but we'd grown distant in the years I'd spent travelling in Central America and living in Uganda. We kept in touch, but without the closeness we'd shared when we were younger. I no longer felt at home in Canada. I missed my home with Akello in Uganda, across from the grandmother tree that fed and sheltered the birds.

My best friend asked me to help her plan her wedding, and I couldn't rise to the task. I became so dishonourable a maid of honour, in fact, that she asked me to step down to be a bridesmaid. "It doesn't seem like you can handle the stress," she said. It stung, but she was right.

I wanted to find short-term or seasonal contract work, which would still allow for some travel back and forth between Canada and Uganda. There was no shortage of contract jobs available to me in Edmonton, but I couldn't seem to hold one down. Through a friend of a friend, I agreed to take on a well-paid craft services position on a film set, making grilled cheese sandwiches and coffee for the film crew, along with the cast of B- and C-list actors. I was on my feet all day, keeping up with the food prep and cooking and making pot after pot of coffee, from dawn until nightfall.

"Where's the hot breakfast?" an actor barked at me one morning, apparently dismayed by the prospect of cold cereal. People barely glanced at me when they approached the food table. I shrank further into a feeling of invisibility, and quit halfway into the contract.

Shortly after, I found another job as a worker on a family-run farm. Every morning, I pulled on paint-stained jeans, an old, faded hoodie, a goose-down vest, and rubber boots, and drove an hour south to a farm outside Edmonton. It was March and the snow was giving way to the coming season. In the fields, the grain stalks shot out of the earth like beard stubble. The song of geese was drifting north again. They landed in the fields and drank from the slushy, semi-frozen dugouts.

I fell in love with labouring on the land.

There was no pretension, no posturing about the tasks at hand: sort through mixed greens, pick out rotten leaves, bag and weigh the greens for market. Or, drop minuscule kale seeds into seeding trays. One seed, two seed, three seed, four thousand seed. I loved working in the greenhouse, feeling the warmth of the lights, the radiant heat, my eyes reawakened by the greenery of kale and mustard and rainbow Swiss chard that grew year-round. I basked in solitude, planting seeds with a covering of fine soil, listening to my favourite folk songs or a *This American Life* podcast.

After a few weeks on the farm, I took over the responsibility of collecting duck eggs every morning. There were specific rules I had to adhere to: Move slowly, in clockwise fashion. Don't scare or startle the ducks. Don't you dare wear red. Ducks, I learned, were highly sensitive to change, the slightest turn in routine could upset the number of eggs they laid. Our quota was around seven hundred duck eggs a day. I'd burrow through warm sawdust with my hands, searching for the large, hot eggs, and fill wire egg baskets. The two hundred ducks quacked and waddled with the energy of a single white, feathery organism. Any sudden movement upset the streaming flow of birds.

Bent over the processing sinks, I scrubbed the duck eggs clean of waxy afterbirth and sawdust. I held them up to the light, examining their mint green and cream and warm sand–coloured shells for any hairline cracks. The work was back-breaking and tedious and thankless, but I knew exactly what I'd accomplished every day. My body grew new muscles. My mind felt swept of worry, scrubbed clean by physical exertion. At lunch, I and the other workers gathered around the table in the family's home and wolfed down plates of roast duck and vegetables dripping with gravy.

The work was enormously satisfying, but at twelve dollars an hour it couldn't pay the bills. Not enough to send back to Uganda, nor enough to hire an immigration lawyer, buy plane tickets, and save for the future. Though I wanted to stay, after a few months of looking around for alternative work, I landed a corporate communications contract position downtown that offered double my hourly wage on the farm. I handed in my two weeks' notice of resignation, and the farmer, burdened by debt and the exhaustion of struggling to keep a small farm alive, was furious with me.

"Don't come back tomorrow," she said hastily.

Heartbroken, I said goodbye to the ducks, the geese in the fields, and traded in my faded jeans and rubber boots for dress pants and shiny black flats. I moved my belongings into a windowless cubicle.

I had been hired to write, but I barely wrote a word. No one seemed to know what my job was about—including me. At least my paycheques were substantial enough that I could both save and send money back to Akello. But in my culture-shocked state, I fared poorly in corporate cubicle land, grazing elbows with my colleagues but barely speaking. I sat in my cubicle longing for the kind of light that coaxes life out of the soil, that makes flowers turn their heads towards the sun. The kind of light that grows hope.

To cope with my loneliness, I burrowed deep into my work of writing a book about the inspiring efforts of women farmers, and pored through interviews I'd collected with female farmers and farm workers. Only when I faced the page did I feel a sense of home within myself; my fingers flew across the keyboard, and the stories began to string themselves together. I was doing what the young, adventurous girl in me had always wanted to do in life—write a book. A literary agent in Toronto agreed to represent me, which strengthened my resolve to finish the project. It was a way to shelter myself from the uncertainty of everything else in my life.

Only a month after I interviewed for the fire tower job, I received a phone call from one of the forest rangers. It was early February and I was staying temporarily at my parents' place in Peace River, unemployed, finishing my book—and hopeful for a job opportunity. My mittened hands pawed awkwardly for the phone.

"We'd like to offer you a position at a fly-in fire tower in the Peace Country," he said.

"Oh," I exhaled. My breath became a frozen cloud in the minus-twenty-something temperature. My surprise hanging before my own eyes. They thought I was capable enough—or crazy enough—to do the job. I had fooled them, somehow.

"Thank you," I said to the ranger, accepting.

Despite the relief I felt, the knowing I'd have a secure income for the summer, I was also afraid. I looked out at the frozen river, at the slabs of layered ice piled haphazardly, locking down the brown body of water that flowed beneath. The armour of ice calmed me. Fire season was far away. For now, the fire tower could remain mythic, a hypothetical entity, as strange and distant as Mars. The river groaned. The slabs of ice creaked and shifted. A large raven perched on the rib cage of a deer carcass. Only bone and hoof remained, and the legs arched towards the sky, as though frozen in flight.

I waited until nine o'clock in the evening to share the news with Akello. I called him over Skype, staring into my laptop screen, waiting for his round, smiling face to appear. In Uganda, the world was waking up. I could imagine the slatted light pouring through the iron bars of the windows in his home. The mammoth banyan tree that towered across the road, teeming with early morning birdsong. Women, already in the compound, hanging wet clothes on the network of wire clothes-lines, strung this way and that. He materialized on the computer screen, perched at our low wooden coffee table, spooning heaps of sugar into a steaming cup of milk tea.

"Good morning, my love!" his voice boomed loudly. "Or should I say, good evening!" A smile spread widely across his lips, and though he was more than fourteen thousand kilometres away, I felt every part of my body smiling too. Even the fine hairs on my forearms seemed happy.

"I got the job," I told him.

"Oh, congratulations! That's great news!"

"Yeah, it will be good money," I said. "But it also means I won't be home to visit for another eight months." The words formed a hard stone in my throat.

For a moment his face turned sombre, then he forced a smile. "It's okay, small one," he said. "Remember, the tortoise is slow but he always gets to where he wants to go."

Always a proverb. When I left Uganda several months earlier, he'd pressed a small wooden carving of a tortoise into my palm, and whispered the same proverb into my ear. "Hold on to hope. We'll be okay."

Akello was always so certain about everything. So hopeful. But he knew less about the immigration and political systems we were up against. That wasn't his fault, but even so, I was beginning to resent these proverbs. I didn't need words of reassurance, I needed tangible help. Someone to fill out the paperwork, organize the files, ready the application. We needed money to make this work, and his job at the garage paid mere shillings, enough to buy a load of groceries, or a meal at the pork joint, but nothing more. My knees were buckling under the weight of pressure to navigate the immigration system in Canada, and he was calmly offering proverbs from our beautiful home in Uganda.

"But let me ask you one thing," he said, more fretfully. "What about the bears?"

I laughed. The bears. The man who'd grown up with black mambas and lions and hyenas was more afraid of bears than the bureaucratic challenges we faced. I was more afraid of the gap widening between us:

continental

drift

My first few months in Uganda had been lonely ones. While I wanted to belong to a community in Kabale, I knew it would take time to develop deeper connections with people. I didn't yet know the culture or the local language, Rukiga. I tried to blend in to my surroundings, even though a part of me would always be the *muzungu*, the foreigner.

I rented the living space above the home of Dr. Adroa, and every morning I walked from my apartment to the health care clinic, where I worked at my laptop in a windowless office. It was a brief stroll to the clinic, but never a boring one.

I passed an open lot where people dumped and burned their garbage. Stray dogs with washboard ribs nosed through the waste. Some mornings there were young men—boys, really—pilfering for food, clothes, anything useful. "*Agaandi, muzungu?*" they called at me. How are you, foreigner? "*Niyge!*" I responded. I'm well! And they laughed because, no doubt, it was funny when a foreigner tried to speak Rukiga.

I passed women scrubbing clothes in soapy water, women peeling plantains, women selling sweet potatoes from woven baskets balanced on their heads, women carrying infants on their backs. Everywhere, women working. My feet kicked up loose orange dust, staining my white socks. A young boy flicked a tree branch on the flank of a long-horned Ankole cow. I passed the business fronts of a general store, a dairy, a cellphone outlet, and a pork joint where a woman deep-fried chips and men sat at plastic tables, sharing the same platter, and using their fingers to spoon down pork and cabbage slaw. People saw me coming and going, every day.

"Hello, doctor!" they waved, unaware that I knew nothing about medicine.

After a month, fewer people were calling out at me "*Muzungu!*" and more were calling me by name. Colleagues and acquaintances stopped to greet me in the streets. "How *are* you?" they asked, gripping my hand warmly with both hands and not letting go. I didn't mind these long-drawn-out greetings. I began to hear myself laughing more frequently. After work, I put down my books and spent the evenings with Patricia, a housekeeper at Dr. Adroa's home, keeping her company as she prepared *karo,* a viscous, doughy bread made from cassava flour and hot water. Her father had died when she was eighteen, and she'd moved from northern Uganda to Kabale several years ago to work for Dr. Adroa and save money to send back home to pay for her younger sister's school fees.

Every night, Patricia and I traded stories back and forth. One evening she laughed to me: "Trina, I wish to paint you black. You are white, but you have a black heart." She had told me stories about the way some of the foreign visitors often made her feel: lesser than, poor, in need of their charity. Many came to Kabale with a superiority complex, or white saviour attitude. They didn't see her strength, her creativity, her humanity. How her heart could expand and break just the same. Patricia and I were very different, and yet we weren't so different—unmarried women in our late twenties, trying to find meaning and connection, a sense of purpose beyond marriage, never short on contradictions, and full of dreams and desire.

Every Sunday, Patricia took me to church. I wasn't religious, but I loved strolling through town with her, sitting together on the pew with a shared Bible balanced across our thighs and the same songs on our lips. "Trina," she said playfully, teasing me. "I think you might have been my long-lost twin."

I hadn't come here to lose myself, I realized. I was learning how to belong.

CHAPTER THREE

In early April 2016, I drove down south to the Hinton Training Centre, a government facility, to attend an eight-day mandatory Lookout Observer training course. The campus consisted of an administration building, two residential buildings including dorms, a mess hall, and a recreation room. There was a building with five flights of stairs for firefighters to launch themselves off and rappel down, and a forty-foot practice fire tower nearby.

In the student residence, the military-style dorm rooms held two single beds, two drawers, two desks, and two lamps. I'd be bunking with a detection aid, a tower support person from the High Level district. Lining the hallway walls were a series of black-and-white photographs of forestry personnel taken over the last century: rangers, foresters, firefighters, pilots, and lookouts. One of the photographs, dating back to the 1950s, portrayed a group of lookouts gathered around a large compass device, the Osborne Fire Finder, which was used to locate the bearings on the smoke from wildfires. I looked into the faces of these men and wondered: Where were all the women? Later, I'd learn that the first female lookout wasn't hired until the early 1970s; apparently, she'd applied for the position using a man's name. Much had changed in forestry over the past fifty years, however, with more women stepping into roles that had been traditionally held by men: rangers, firefighters, helicopter pilots, air-attack officers, and so on. Recently, I was surprised to learn, more than half of Alberta's lookouts were women.

The following morning, the cafeteria was crawling with firefighters: young, athletic men and women—although mostly men—dressed in the standard mustard-yellow-and-green Alberta government uniform. The rangers, men and women in their late thirties and forties, wore sand-coloured shirts and slurped coffee at their own tables. The radio dispatchers sat at a table by the window—all women, I noticed, who appeared to be in their early to mid-twenties, and who looked, on the surface, a lot like me. I began to wonder if I had applied for the right job. Maybe I'd be better suited for a summer working the radios. At another table sat the dozer bosses, the guys who operated bulldozers, responsible for clearing the bush down to mineral soil to prevent the spread of larger, more volatile wildfires. They had a harder look to them, with grizzled cheeks and paunchy bellies. They wore blue jeans and truckers' hats and loitered around the entrances sucking on cigarettes. I figured that the attendees dressed in plain clothes, who didn't look as though they belonged to any particular group, must be here for the training to become lookout observers.

There were only a handful of us. We were a motley mix of candidates, about a fifty-fifty split between men and women, and a wide range of ages, from mid-twenties to mid-sixties, including a couple of forestry students, a musician, another writer, a biologist, a photographer, a retired teacher, and a former trapper. We were a pack of quirky, introverted nature enthusiasts, each of us thrilled to have landed one of the few openings. Of the 127 fire towers in the province, only 10 had required new lookouts for the 2016 season, which, I'd learn, wasn't at all uncommon. The majority of lookouts tended to migrate back, summer after summer, to their towers. Many were what they called "lifers," operating on a seasonal lifestyle, building second lives and homes at the towers.

"It takes at least five years to make a good lookout," one of the managers told us on the first day of training. "What we teach you here this week, you'll have to apply at your own tower."

No two towers were exactly the same, he said. There were towers situated atop mountains, like Jack Kerouac's refuge on Mount Desolation, a single-room structure in which the lookout both lived and worked, looking out the windows for smoke. Other alpine lookouts were two-storey structures with a cabin on the main floor and a ladder or stairs that led up to the cupola. There were 40-foot towers, 60-foot towers, 80-foot towers, and 100-foot towers—and even a few rare 120-foot towers. There were towers in the foothills, overlooking the southern wild grasslands; towers surrounded by deciduous forest; towers that peered into the fields of farmers and ranchers, or ones hunkered down next to oil well sites and gas plants. There were towers next door to military aerial operations, and ones you could reach only by helicopter. Towers that would only ever see fires caused by lightning, and settlement towers, which rubbed up against the edges of communities where humans usually caused fires. Some towers were so far north that they sat on the Canadian Shield. The farther north you travelled, the more the black spruce shrank down to stubble. I would be going to one of the last true "wilderness" towers, inaccessible by road, a tower overlooking an expanse of forest.

Over the next week, we sat through lectures on weather, fire behaviour, and occupational health and safety. We studied photographs of smokes and false smokes—phenomena such as road dust and pollen clouds that look deceptively like smoke. The trainer flipped through slides. "Smoke, or false smoke?" he asked. My eye zeroed in on the column of black smoke rising from the stand of pine and spruce. "Smoke," I said confidently. "Look again," he said, and I immediately noticed the black cylinder beneath the smoke column. An industrial flare stack. There were a thousand ways to be fooled from the fire tower, particularly from forty kilometres away. Even a tuft of cumulus cloud, rising behind a distant ridge, could look like a puff of white smoke. Sunlight falling on a distant cutline, or a stand of white birch or aspen. Fog draping over the sharp teeth of the spruce, or columns of moisture rising after a violent storm. A ribbon of black diesel coughing out of the exhaust pipe of an old tractor and staining the sky.

"It's not necessarily about just looking and seeing out there," said one of the managers. "It's more about knowing where, when, and how to see."

Before training began, I wondered if the government trainers would bring up the disappearance of Stephanie Stewart, now a decade-old unsolved mystery. I was curious if my fellow trainees were even aware that a lookout had gone missing, that she'd never been found. The story lingered in the back of my mind. It was a reminder of a risk, a reality: I was signing myself up to live alone in the woods. I would be vulnerable.

"One of our lookouts was murdered in 2006," the head manager said solemnly on day one.

My eyes widened. I felt my colleagues stiffen and suck in their breaths. That word—*murder*. My body felt cold, the kind of damp cold that sinks into your bones. The RCMP had recently reclassified Stephanie's case from "missing" to "homicide." There had initially been rumours of a bear attack, or speculation that she'd wandered alone into the woods, but they'd quickly ruled out those possibilities. Very little information had been published in the media after the disappearance, but apparently the RCMP had enough evidence to be certain of foul play. I respected the trainers for addressing the event directly, though it only fuelled the fear I already felt. The memory of my own trauma reverberated in my body. I wondered whether symptoms of my latent PTSD would emerge at the tower.

On the third day, we practised locating and mapping wildfires using the Osborne Fire Finder, the same device that the group of lookouts had gathered around in the photograph dating back to the 1950s. The Fire Finder was a large, heavy compass etched with bearings from 0 to 359 and set up facing true north. A rifle scope, a telescopic sight, was attached to it. When it was my turn, I swung it around, focusing on the mock smoke through the crosshairs, and then looked down to record the corresponding compass bearing—56 degrees and 10 minutes. The mock fire was ten kilometres away. I then rushed over to the cross-shot map, a map with a network of large circles indicating fire towers, and pulled a piece of string from the centre of the map—the location of my fire tower—along the compass bearing to a distance of ten kilometres. The next step was to call my hypothetical tower neighbour. "What's your bearing on the smoke?" I asked through the hand-held radio device. "Two two-five degrees," came the response. On the map, using a second piece of string from the neighbouring fire tower, I triangulated our bearings and pinpointed an exact location. There. X marks the spot.

It's a fact that a compass, a map, push-pins, and two pieces of string have been the trusty tools used to detect wildfires for over a century in Alberta. Of course, there are other methods: helicopter patrols, fixed-wing aircraft patrols, 310 calls from citizens who spot fires along roads and highways and train tracks. But nothing can replace a set of seasoned eyes able to differentiate between a real smoke and a false one. The most experienced lookouts in the province could estimate the location of a wildfire—sometimes burning fifty, even sixty kilometres away—within hundreds of metres from where it burned. It wasn't a matter of seeing so much as knowing the forest and features around the forty-kilometre radius that was each lookout's responsibility.

We'd have to know every road, every cutline through the bush, every set of rail tracks, every farmhouse, every lake, and every ridgeline. Some of these features we'd be able to see from the tower with the naked eye, or through binoculars, but others would be blind to us, hidden behind hills and dense forest.

"You'll learn to keep extra vigilant on long weekends for campfires," they told us. "And on windy days, watch for permits jumping into the bush." This last bit referred to a farmer's brush pile fire accidentally blowing into the forest.

There were dozens of scientific names for clouds that would help us understand the weather and risk for fire hazard in the forest. Cirrostratus, altostratus, altocumulus, cumulonimbus—I couldn't keep track of them all. The cumulonimbus, or thunderhead, was one to stay alert for. A ranger flipped through photographs showing the progression of thunderstorm development, the first image showing a mere whisper of cloud. The barely visible cloud grew steadily into a dark tower, its head becoming a monstrous, angry, mutant cauliflower. These were the kinds of clouds that would shoot lightning strikes capable of starting fires.

Lightning caused fewer than 50 percent of the wildfires in Alberta, but where these fires ignited, I learned, they burned the largest areas of forest in the province. The tower I would be going to—in less than a month—was considered a lightning tower because the surrounding area was mostly uninhabited by people. Fires in my territory were much more likely to be caused by natural phenomena than by humans.

"Don't climb your tower if there's a storm cell within twelve kilometres," said one of the rangers. They assured us that the towers had lightning grounding systems, underground systems to absorb the shock of a strike, but *only* if we were tucked safely in our cupolas, or cabins. Recently in the United States a woman had died when lightning struck her tower while she was climbing the steel ladder.

Over the course of the training, the instructors grilled many "don'ts" into us.

Don't climb the tower if there's more than ninety-kilometre-per-hour winds.

Don't expect your food to show up exactly when scheduled.

Don't leave your garbage outside, because it could attract bears.

Don't drink unfiltered rainwater without boiling it.

Don't say anyone's name over the radio, for security reasons.

Don't go for a walk without your radio and bear spray, and without telling a neighbour of your whereabouts.

Don't expect minor issues to be resolved at the drop of a hat.
And whatever you do, do not go out there to find yourself.

Akello and I first met on a Sunday while washing our clothes outside in plastic basins at Dr. Adroa's compound. The doctor was his uncle. I had noticed Akello around before that day. He was tall and slender, and much darker skinned than many of the men in Kabale. He had grown up in Arua, a province in northern Uganda bordering South Sudan, and belonged to the Lugbara ethnic group. He had an enormous laugh, one that would make anyone laugh just by hearing it, even if they had missed the joke. His laugh was made of pure joy, and it ricocheted off the hot concrete as Patricia and I did laundry.

I dumped my clothes into the sudsy water basin and awkwardly used my arms to replicate a spin cycle. Truthfully, I had never washed my clothes by hand before.

"Is that how you wash your clothes in Canada?" he teased.

"Well, no," I said self-consciously, suddenly feeling like a child splashing around in the water.

Without a word, he reached his long arm into the basin, fished out a black bra, and took the bar of detergent from my hand. I watched, mortified, as he started kneading one bra cup into the other, scrubbing the fabric clean. My cheeks flushed hot. He continued chatting, unperturbed. Panicked, I reached into the soapy water to grab my underwear before he could wash those too.

He was kind and playful, and not shy. I liked him. I could tell because my laugh grew new shoots, sudden abundance. Patricia looked up from her basin and shot me a curious look. He asked about the tattoos on my arms, a Cuban moth on my left wrist and the watercolour flowers that cascaded down my right forearm, a tattoo I'd had done in Guatemala years ago. He told me he was studying mechanical engineering at a local college in Kabale but that he felt happiest on the football pitch, dancing circles around the other players. "Or 'soccer,' as you North Americans say," he laughed. He'd grown up in the north but spent the majority of his youth in southwestern Uganda, growing cassava and coffee on his grandparents' farm. He was what people in southwestern Uganda would call an *abataka*, a man of the soil.

I loved the way his name rolled off my tongue.

"What does your name mean?"

"I cannot say," he told me, smiling widely, scrubbing an orange-stained sock white again. "It's a sad story."

I couldn't tell whether he was serious or joking, so later I asked Patricia, who was also Lugbara, and had grown up with Akello in the same town in northern Uganda.

"It's Lugbara for 'no father,'" she told me. My face turned solemn, so she laughed, trying to lighten the mood. "But we used to call him Ake for short, which means 'bald.' He had no hair when he was younger! We used to tease him for that! Everyone loved Akello. He was a good worker in the fields. He always helped his grandmother with the digging."

In Lugbara culture, she explained, babies are named after circumstances around the time of the birth. A baby may be named Drought or Famine or Jealousy—if someone in the community was jealous of the mother, said Patricia—or even No Father.

Akello's happiness seemed infinite, uncontained. I wanted to know about the texture of that laugh, to hold that joy in my hands. I dreamt of those large, kind hands enclosing my own.

On the fifth day of training, we boarded a school bus to drive out to a nearby fire tower for an orientation and practice climb. I sat beside a handsome, shaggy-haired man named Sam. Now in his late twenties, he had worked the previous summer at a lookout in the Lac La Biche district, but he'd been hired on the spot—thrust into the job last minute—and had not attended the training. Like the old-timey lookouts, he'd learned on the job. No mock fire simulations. No practice climbs. They'd just dropped him off at his tower with a climbing harness, a training manual, and all of his belongings. This year he'd be working at one of the northernmost towers in the province.

"No industry," he said with relief. He had been surrounded by oil and gas well sites at his previous tower. But the tower he was headed to this year was a fly-in, surrounded by nothing but scraggly black spruce, muskeg, and views of the big smokes on the horizon, fires allowed to burn freely in remote, uninhabited areas of the Northwest Territories. It was just what Sam wanted: to be left alone, to hear nothing but wind in the trees, birdsong, and the music of his own making. Some towers, especially those closer to human dwellings and industry infrastructure, were considered fire hot spots, with lookouts reporting twenty, thirty, even fifty smokes in a single fire season. Other towers, particularly the alpine lookouts in the Rocky Mountains, received over three thousand visitors—hikers, hunters, recreationalists, Girl Guide groups—every season. A wildfire ranger would later tell me a story about one of the alpine lookouts having to interrupt a group of yogis who were congregated on his concrete helipad, their bodies bent in Downward Dogs and rising up in warrior position. So much for solitude. But where Sam was going was on the edge of the map. No one from the public would ever visit, let alone lay eyes on his fire tower. The only visitors he'd receive would be forestry staff and lynx and black bears. The fires he'd report would burn vastly, fiercely, chewing up old forest, but they'd never be known to people beyond pilots and firefighters. They'd rarely make the news.

"What was it like out there in your first season?" I asked Sam.

"Pretty surreal," he said serenely. "Most days, anyway."

Sam told me stories about harvesting wild camomile and mint and setting up an old bathtub overtop his firepit. It was all so romantic. What he didn't talk about—not until much later in our friendship—was the emotional agony of trying to support his ex-girlfriend, who was experiencing suicidal thoughts, from the distance of the fire tower, of feeling trapped by the isolation and unable to physically help someone he cared about.

He was from Halifax, a recent political science graduate, and a former student politician who had burned out from the demands of political activism and social organizing. Sam was unsure of his next step. Like me, he had backpacked through Central America and parts of Europe. I told him about the book I had just written about women farmers. Oddly, we discovered that we had both travelled to the same tiny, off-the-map village in northwestern Guatemala, a community that no tourist would ever visit, not unless they were somehow connected to one of the local organizations that were actively protesting against a Canadian-owned gold mining operation. I spent several weeks there, learning from the Mayan-Mam farmers about their century-old ancestral maize and the fight to close the mine.

"Were you there because of the mining?" he asked, peering a little closer at me.

"Yes," I said in a single breath, incredulous. "You too? What are the chances?"

"That's bizarre." He laughed. I felt an immediate kinship with him, a little rush of pleasure to have discovered Sam, an ally perhaps. I hadn't expected to make any new friends as a lookout. I'd anticipated only my solitude. Even so, I didn't tell Sam about Akello. I didn't tell anyone about Uganda, for that matter. I often struggled to open up to people in Canada about my life in Uganda and my decision to marry Akello. I didn't like to feel as though my life was somehow different, or exotic. Sometimes the way people looked at me, particularly after I told them about the process of immigration, felt condescending, or patronizing. "Well, I wouldn't want to trade places with you," a friend once told me. And sometimes, although they wouldn't say it, I sensed their judgment that our relationship was doomed to fail.

The bus pulled up to a forestry base: several mobile trailers, picnic tables, an airstrip. We piled into the back of a medium-sized helicopter and buckled up as the pilot's voice came on the radio through the headsets. "Woohoo!" he hollered like a cowboy. I noticed a Canada–Afghanistan sticker on the back of his helmet, and wondered if he had flown for the military. The motor blades roared and we lifted off. My stomach flipped like an acrobat. "Which of you are gonna suntan up on the cupola this summer?" the pilot laughed, and told us a story about flying over a tower years ago. The lookout, a woman, had crawled on top of her cupola, one hundred feet up in the air, to sunbathe. "I caught her in the nude!" he howled with laughter.

I wondered if his story was an exaggeration of the truth. I was starting to question some of the stories I'd heard previously about the "exotic" lookouts. I'd come to the training with an expectation that my colleagues would be over-the-top-eccentric individuals, but everyone seemed so normal and level-headed. Introverted, certainly. Passionate about their chosen hobbies, yes, but about to crawl through the hatch of their cupola and bare it all for passing air traffic? I glanced around at my new colleagues. Probably not.

I smiled nervously at Sam. I was petrified of even *climbing* the tower, let alone balancing atop the cupola roof. That suntanning, risk-taking woman would never be me.

Our destination fire tower was situated atop a small hill overlooking a gravel-pit operation. Hardly the image of pristine forest. But this was Alberta: big oil, big forestry, big mining. The helicopter landed on the far side of the tower clearing. We ducked out one by one, moving away in a single column from the danger of the churning blades.

A ranger led us around the tower site. I peered inside the 200-square-foot cabin. It was spartan, nondescript, undecorated. It had a gas stove and oven, a kitchen sink—but no running water—a propane-run refrigerator, and cupboards with the bare cooking essentials. On a small desk sat a telephone, a two-way radio, and stationery supplies. The bedroom held a single mattress and a closet, and someone had constructed a narrow shower stall in behind the bedroom door. Here, a lookout could hang up a solar water bag for a quick rinse. The cabin was comfortable, far less rustic than I'd imagined, but had a sterile feeling to it. Standard government model. A shell to crawl into and call, temporarily, home. A path of concrete sidewalk blocks led to a small outhouse.

The ranger showed us inside the engine shed, a small tin building adjacent to the cabin. The shed housed a Honda generator, the only power source, for supplying power to the radio and telephone and three to five hours of electricity a day. The generator ran on propane from a 500-gallon cylinder that had to be trucked in. Helicopters would sling in smaller, 250-gallon pigs to the fly-in towers. Rain barrels would collect water for bathing, laundry, and washing dishes.

The steel tower sat on the tallest point of the hill. One by one, like ants, we climbed the 100-foot fire tower. I watched my colleagues struggle into their canvas climbing harnesses and clip into the fall arrest safety system. Since the early twentieth century, lookouts used to free-climb without harnesses, but in 2000 a lookout in northern Alberta had suffered a heart attack and fell to his death. The following season the government installed fall arrest systems on all towers, and climbing harnesses became mandatory.

I slipped the harness over my shoulders and reached for the long straps that dangled between my legs. The harness sagged low, several sizes too big for my 115-pound, five-foot-three-inch frame.

"Get your supervisor to order you an extra-small," advised the ranger.

I felt childish, birdlike, in the large harness. My heart hammered in my chest. What had I expected? I had signed myself up for this very moment.

Strange that I'd hardly given any thought to actually climbing the tower. I stared straight up at the cupola, a small octagonal box in the sky, a faraway place that required an absurd amount of effort—and overcoming my fear—in order to reach. Heights don't freak me out. I'd always been a fan of hiking and scrambling up steep mountain pitches. In Guatemala, I'd launched myself off a forty-foot bridge into a river. In Nicaragua, I plunged thirty feet down into a narrow canyon opening and hit the water so hard that I nearly dislocated my shoulder. I loved the sensation of looking down, anticipating a jump. But ladders were another story. I hated the feeling of clinging to a ladder. Years before, I'd worked for a student painting company and had to hold my breath on the rickety ladders, even though I was barely fifteen feet off the ground.

The westerly wind was blowing in from the Rocky Mountains and the frigid air stung the exposed skin on my hands. I white-knuckle-gripped the steel rung of the ladder.

"Use the three-point system!" the ranger yelled up at me.

Slowly I ascended, ten feet, then twenty, with three limbs on the ladder at all times. Left hand up, right hand up, left foot up, right foot up. Repeat a hundred times.

At fifty feet, I glanced down and wanted to puke. What the *fuck* was I doing? The wind whipped the hair around my face. Fear seized my forearms. It was the wrong way to climb, I knew, holding all the tension in my arms. The power should come from my legs. But fear has a way of gripping your body, and thus I squeezed the ladder rungs—a death grip. Just keep climbing, just keep climbing, I whispered to myself through clenched teeth.

At eighty feet, I wouldn't let myself look down. The wind whistled loudly through the guy wires of the tower. I felt an ocean of space around me. Instead of down, I looked up at the underbelly of the cupola. The hatch door was so close, but so far. My forearms ached. Pride was the only thing that propelled me upwards. Finally, my heart pounding in my ears, I pushed my skull against the bottom of the hatch and popped up into the cupola.

I hoisted myself up onto the fibreglass floor, shut the hatch, and unclipped myself from the harness. My lungs were working like over-tired moth wings, my heart slamming in my ears. A ranger offered me a high-five. "Good job!" she said, smiling widely.

Immediately, my eyes were drawn to the window, the view travelling far towards the mountains rising like jagged molars in the distance. The cerulean sky sang like a promise. Aside from flying in a plane, I had never been so close to the clouds before. It felt as though I could reach out and run a hand through them. The cupola rattled back and forth in the wind and my legs braced in the sway. Somehow, being above the world made everything incredibly dazzling. I watched the back of a raven kiting by in a wind draft. For just a split second, the sun turned the raven a shade of indigo. A shot of fleeting beauty.

Maybe I could do this after all.

On one of the final training days, we moved the desks and chairs in the classroom into a circle. "This is a safe space," said the man with a neck as thick as the trunk of a mature spruce. He was a self-defence instructor and judo master. The veins on his forearms bulged.

The idea of a mandatory self-defence course immediately set me on edge. The interview question about the two hunters showing up at the cabin rang in my ears. My body couldn't forget what had happened in Nicaragua, being trapped in darkness, the man's hand over my eyes, the fists pummelling my chest. I wondered how my colleagues felt. Everyone looked a bit unsure, awkward, gathered in a circle, avoiding eye contact with one another.

"The goal is to practise de-escalation," said the judo master. "If there's an angry guy at a bar, or someone aggressive shows up to your tower, how can you help to calm them down?"

Practise open body language and posturing, he answered for us. Crossed arms would signal aggression, he said, potentially escalating the predator's anger. Open arms signal "let's talk."

"Let's practise," he said, motioning for us to split up into pairs. "One of you play the predator, the other the victim." The words *predator* and *victim* left a strange taste in my mouth.

The class erupted into a cacophony of mock-angry voices. My partner, nearly six feet tall, rushed towards me. Even though it was an exercise, I winced, feeling a surge of anger. If someone showed up at my tower with a premeditated plan to harm me, what difference would my arm position make?

My PTSD was aflame.

Abruptly, I walked out of the room, down the hallway, and into the women's washroom. I sat down on the toilet and my eyes welled up with tears. Why was I thrusting myself into such an uncertain, risky situation? Based on my own history, was I even emotionally equipped to handle the isolation?

Later, I debriefed these questions with one of the trainers, a woman who'd been working as a lookout at a remote boreal fire tower for five years. She listened sympathetically, and told me a story about having to deal with a group of male mushroom pickers who randomly showed up at her tower one day and, uninvited, started to climb the ladder. "I stood on the cupola hatch door and yelled down at them to leave," she told me. "They eventually got the hint."

I nodded, my fear softening, instead feeling something of awe for this strong, self-sufficient woman, probably the same age as me, who'd become such a well-respected lookout in Alberta. I appreciated her ability to relate to my fear of the unknown and worst-case scenarios.

I wiped my face and followed her back into the hallway. She offered to teach me the self-defence moves herself.

"Don't overthink it," she told me gently. "You're going to do great out there."

Later that evening, Sam and I hiked down to the railroad tracks that run through Hinton. We sat on the grassy slope beside the tracks, lay on our backs, and looked up at the stars.

I opened up to Sam about my emotional reaction to the self-defence workshop, explaining what happened to me nearly six years earlier in Nicaragua. Those two words, *what happened*, a euphemism for the event that changed everything right down to a molecular level inside my body.

I told him about the assault and the string of events that happened in the months immediately after: Calling 911 in the middle of the night while house-sitting for my parents, when I mistook the sound of their cat coming down the hallway for an intruder. Or, worst, the night I was out drinking with friends and picked a fight with the guy ahead of us in a lineup who'd loudly joked about the woman in front of him. "I'd do her in the ass," he'd said for everyone, including the woman, to hear. I'd thrown my half-eaten pizza on his expensive Italian leather shoes. "Do you know what a *misogynist* is?" I'd challenged him. I confided in Sam about being diagnosed with PTSD and seeing a psychotherapist in Edmonton to address the trauma.

"I'm so sorry," said Sam. When he hugged me, I felt safe against his warm chest. Then I felt a pang of remorse, as though I was somehow being unfaithful to Akello because I wanted to feel safe, to feel loved, to be touched. Across the distance, I felt our connection beginning to dissolve.

Sam and I watched the stars for a while longer, not speaking a word. I anticipated the wordlessness that was to come. Despite my fear, I was comforted by knowing that Sam would be out there facing some of the same challenges as me. I was grateful to have met him face to face, someone I felt I could trust—an unexpected friend.

After the training wrapped up, one of the rangers invited a few of us up to a nearby lookout that had already opened for the fire season. It was the same lookout from which Stephanie Stewart had gone missing ten years earlier. The ranger said he came back every year since she'd disappeared to pay respects to her memory. These days, it was staffed by another veteran lookout named Jack.

We drove along the muddy gravel road, accelerating up and down the side of a mountain, towards a clearing where the cabin perched atop a rocky peak. The cabin was gorgeous, hand-constructed using spruce logs, with a huge wraparound porch. Jack, a grey-haired man in his sixties, appeared on the porch and waved us in for a visit.

A forty-foot tower stood above the cabin on the highest point of the yard, overlooking the forest-covered mountains and the town of Hinton nestled down in the valley. Jack seemed happy about the visit, and even happier when he discovered that we'd all soon be flying off to different lookouts for the summer. He eagerly launched into recounting his favourite stories from his twenty-five years on the job. He was what forestry folks call a lifer.

Jack told us about the time he wanted to mess with his junior ranger, a "city boy who'd come to work in the bush," so he took off all his clothes and was spinning doughnuts on his ATV when the young city ranger pulled up to his cabin with groceries. "He was mortified!" he said, laughing. "He thought he was dealing with someone who'd gone totally nuts!" We all laughed.

But when Jack talked about the hard days, his expression grew sombre.

"After the fifth or sixth day of rain, when you're stuck in the cabin alone," he said reflectively, rubbing his grizzled chin, "that's when you start to look in the mirror and wonder, Who is this person?" He glanced knowingly at the ranger and they shared a laugh.

"Some days I climb the tower at night to watch the stars," he said. "You'd think after a fourteen-hour day in the tower looking for smoke it would be the last place I'd want to be. But I don't have to report shooting stars." He smiled. "They're just for me."

Akello agreed to help me plant a garden on the sloping land behind the doctor's house. We chose a small plot beneath the shade cast by a cluster of banana trees. He handed me an *efuka*, a hand hoe with a long wooden handle and a heavy, broad blade, then took off his shoes and socks, rolled up his pant legs, and slipped barefoot into the soil until he was ankle deep in the earth. He showed me how to swing the hand hoe high above my head and allow gravity to fell the blade into the soil. His knowledge and ease with the earth drew me closer to him.

We planted corn and beans and wild pumpkin seeds, cast handfuls of carrot seeds and dusted them with soil. We pushed *Sukuma-wiki*, a kale variety from Kenya, into the loose, turned earth. Akello built a nursery out of woven banana leaves, felling the tree with a single swipe of the *panga*, a large machete. Beneath the nursery stand I planted basil seeds that I'd smuggled from Canada, hidden in rolled-up socks in my suitcase. I wanted to learn from Akello but also plant something familiar, something of myself, too.

We would meet and tend the garden nearly every day after I finished work. I would hear him before I saw him, singing an upbeat Congolese song, or laughing, while pushing his bicycle, coming back from class at college. He could see possibility before I could, the slightest fleck of green emerging, pushing up through the soil. He taught me how to see in the garden, pointing out the wild things that grew—passion fruit and avocado seedlings—that we shouldn't weed. He told me the names of birds I'd never seen before: crested bulbuls, yellow-breasted weavers, and a blue heron that became so comfortable with our presence in the garden that he appeared silently, stalking snakes and frogs that sought shade beneath the broad-leafed beans and towering corn crops.

"Trina and Akello love their garden," Patricia said, clucking her tongue, watching us.

During my first year in Uganda, my parents came to visit, and they too fell in love with the community I was cultivating in Kabale.

"I'm not ready to come home," I told them before they left.

My mother smiled through watery eyes. "We know."

I delayed my flight for another three months and declined admission to grad school in Norway.

I told Akello, early in our friendship, about being assaulted in Nicaragua and my symptoms of PTSD. I wasn't afraid to tell him because I knew he was strong enough for such a story. He'd witnessed a lot as a child, after his own mother struggled with mental illness and postpartum depression. He and his siblings were sent to southern Uganda, just north of Kabale, to be raised by his grandparents.

Falling in love with him was as easy as breathing.

After four months, we plucked ears of corn from the garden. In the adobe kitchen outside Dr. Adroa's home, we crouched low around a charcoal cooking pot and grilled the corn over the blistering hot coals until the kernels singed black. We ate the corn, plain, without butter or salt. It was the sweetest thing I'd ever tasted.

Late one evening, as I stood up to say goodbye, Akello grabbed my hand.

"You're going to leave me one day," he said softly, knowingly.

Now that his joy had found me, I couldn't contemplate life without him. I'd never felt that as deeply, as genuinely, with anyone else in my life. I needed him the way my lungs needed oxygen.

"Never," I said. "*Alemisaru, ma pati.*"

I love you, my tree.

"*Ma o'la*," he said.

I love you like food.

CHAPTER FOUR

The snow never came that winter, in the months before my first season at the tower. The Peace River was frozen solid only for a few weeks. The snow atop the valley had melted down early, leaving the half-frozen earth exposed. My father was supposed to fly over the forest, chasing and counting caribou, but biologists relied on the snow to make their counts. Without snow, the caribou blended into the neutral hues of greys and greens, hidden in stands of spruce and pine, embodying their namesake—the grey ghosts. My father and his team were landlocked for most of the winter, counting, instead of caribou, the number of days of cancelled flights.

By the end of March, hardly any snow remained. I imagined the melted snow pooling in the dens of black bears in the river valley, the sows and their tiny cubs stirring from their torpor. One afternoon I heard the faint honking song of Canada geese. "Did you even fly south?" I whispered, cranking my head skyward and watching their black silhouettes beating against the sky.

These sudden shifts stirred something deep, instinctive, and unsettling in me. Sociologists have a term for that now: anticipatory grief, a feeling of sadness in the face of impending environmental loss. Everything was out of whack. Climate change had never felt more real to me. The mountains of snow that once accumulated on every street in town were half-melted rivers in the gutters. The forests drank what they could from the small offerings of snow. The sun beat down months too early and licked up the last of the ice. Signs of green up—the opening of deciduous buds, grass re-emerging from dormancy—were already apparent.

In early April, only a week after training, the fire season was born, a month earlier than usual. I received an urgent email from my supervisor in Peace River, addressed to the fourteen lookouts in our district.

"Everything is ahead of schedule this year," he wrote. "We're opening towers two weeks earlier than planned. Your orientation dates have all been bumped up."

My departure date was less than a week away.

I panicked. There was so much to do still, shopping and sorting and packing, not to mention so many loose ends to tie up on the immigration front with Akello.

I drove down to Edmonton to stock up on dried and canned foods that would sustain me for four months. Food up north cost significantly more than in the city. I pushed an extra-large shopping cart up and down the aisles of a Superstore and began to load up the cart like a pack mule, checking off items on my list:

- ✓ 4 kg dried pinto beans
- ✓ 3 kg rice
- ✓ 1 kg quinoa
- ✓ 4 lbs whole wheat flour
- ✓ 10 bags coffee beans
- ✓ 12 cans whole tomatoes
- ✓ 6 bags penne pasta
- ✓ 50 granola bars

- ✓ 2 kg almonds
- ✓ 2 kg dried apricots
- ✓ 1 kg raisins
- ✓ Large jar of peanut butter
- ✓ Family size iced tea powder

The cart grew heavy and the back wheel squeaked awkwardly under the heaping bulk. A Superstore employee raised an eyebrow and asked, "Shopping for the apocalypse, are we?" At the till, the cashier rang through item after item. "You win a turkey!" she beamed after tallying the total cost.

Back in Peace River, my father drove me out to the gun range. He stapled a large poster of a grizzly bear up on a wooden frame at the end of one of the shooting ranges. The bear's head was down, ears alert. The front leg was extended, as if the bear was walking towards us. My dad paced about six feet back.

"Here," he said. "This is where you want to be when you use your bear spray. Make sure you're not spraying into the wind. Aim low into the bear's face."

Lifting the canister of bear spray with both hands, I pressed the release lever down. A shot of pinkish-orange substance flew out and my nostrils stung. My father was a big believer in bear spray. Studies show that, in human–bear encounters, bear spray is often more effective than a gun, and most of the time safely resolves the situation—without a dead human or a dead bear. In the vast majority of conflict situations between humans and bears, it's the bear that fares worse.

"But," he said, "you should have the gun as a backup."

We paced another four feet back and he handed me the 12-gauge shotgun.

"If the bear is charging you, you'll only have seconds to react, so you have to be fast."

He told me to aim for the chest. The shotgun slug should penetrate the bear in the chest, so the heart would fill up the lungs with blood. But, he warned, even if you hit a charging bear, it might not slow it down. Momentum and adrenalin could keep the four-hundred-pound predator coming fast.

Bears often knock their prey to the ground. For the best chance of survival, my dad said, flip over onto your stomach to protect your vital organs and cover your head and neck with your hands. Bears will often go for the neck or scalp. With a defensive grizzly bear attack—a sow protecting her cubs, for example—experts say the best strategy is to play dead, so as not to pose a threat to her. But if it's a black bear, a predatory attack, the rule is to fight back. Fight with everything you've got.

There were other predators to watch out for too. Only a year before, a young female biologist working in the Grande Prairie region had been attacked by a cougar. The animal pounced on her when she left her tent to go pee in the middle of the night, grabbed her scalp in its jaws, and dragged her into the bush. The woman's colleagues heard her screams and came rushing to help, fighting the cougar off with a shovel. She survived the attack, but just barely, requiring multiple stitches to reattach her scalp to her head.

"If you're attacked by a cougar," a biologist at the tower training in Hinton had told us, "fight for your life. Jam your thumbs into its eyeballs."

I balked at his words, struggling to imagine myself taking a cougar head-on.

Bear and cougar attacks weren't at all common, but they did happen. I'd heard about a forester working up in the High Level region, just north of where I would spend the summer, who had been attacked by a black bear. It had been a predatory attack, not a surprise encounter. She had been standing in a field, her head down, scribbling into a notebook, when the bear came at her and knocked her down. Though it's hard to imagine such a large creature barely making a sound, she hadn't seen or heard the bear coming. Years later, I'd meet the helicopter pilot who had been there with her. When he heard her cries, he came running and helped to fight the bear off.

"You'll learn to develop something like a sixth sense out there," they told us at training. "But you need to be constantly vigilant. Don't let your guard down. And don't even think of putting in earbuds while you're doing yard work or going for a walk."

My father helped me into position. Widened stance. Warrior pose. The 12-gauge was a Defender, a gun built to be lightweight, ideal for defence against surprise encounters. Even so, the gun felt heavy in my extended arms. I awkwardly pulled back on the action, loading a shotgun slug into the chamber. I pressed the butt of the gun up to my right shoulder, resting it against my cheek. Clicked off the safety. Lined up the shot, focusing my eye low to the bear's chest.

"Breathe in and take the shot on the exhale," instructed my dad.

One, two—BOOM!

The gun bucked. My heart went galloping away into the bush. I looked up and saw that the slug had shredded the poster paper. A clean shot through the bear's lungs.

"Good," said my dad. "Now again. And this time—do it faster."

A week before flying out to the fire tower, I made a last-minute decision. I couldn't do this alone. Packing a gun wasn't enough. I wanted another line of defence. A sidekick. A protector.

I began to look for a dog.

One of the trainers, a lifer—with years of experience living close to the land—had asked me, "Are you bringing a dog with you?" He'd looked closely at me. "I wouldn't go out there without a dog."

His words haunted me. If *he*, of all people, a man who could survive alone in the woods with just the clothes on his back, needed a dog, then I would be lost without one.

I needed a loaner dog. An SPCA rescue. A fire-tower-only dog. I wasn't looking to fall in love with the animal or develop a relationship that lasted beyond the parameters of my tower contract. At the end of the summer, I'd be flying back to Uganda.

Furthermore, three years of living in sub-Saharan Africa had hardened my heart **against** the pet culture of North America. In Uganda, dogs and cats served strictly utilitarian functions. Cats hunted for rats and snakes. Dogs were mostly locked up during the day, but roamed the perimeters of compounds at night, barking to alert security guards to the presence of any intruders. My Ugandan friends were genuinely afraid of dogs. "Trina, don't touch it!" they'd scold, and slap my hand away, as if I were a child touching a hot stove. "They are so *dirty*." Akello's uncle kept a long-legged golden mutt named Tiger locked up in a wooden cage during the day, so we only saw him at dusk. He was regularly watered and fed, usually rank-smelling leftover parts from the butcher. But no one fawned or swooned over Tiger. He served a purpose: to defend people and property. No one would stoop so low as to get their love from a flea-infested dog.

I wanted a tower dog with a mean bark to alert me to unwanted intruders: bears, wolves, cougars, maybe even the two-legged kind of predator. I wanted another set of eyes and ears scanning the perimeter, and the kind of nose equipped to smell up to 100,000 times better than a human one.

I went to the SPCA in Peace River and walked up and down the aisles of the dog enclosures. I passed by a terrier mongrel, spinning and yapping shrilly. Too small, I thought. Bear bait. There was an old mutt that walked with a limp. Too slow, I thought. I'd heard stories of wolf packs going after tower dogs. I needed a dog that could run, or fight for its life—and my life too. I stopped by the enclosure of a husky mix. A decent-sized, stocky dog, cream-coloured with blue-and-black eyes. I took Camille for a walk along the river. She pulled, headstrong, as I knew huskies are bred to be. I began to worry: What if she only caused trouble for me? I'd been warned about the risk of bringing an untrained dog. There were stories of dogs going after bears, then bears turning around, giving chase, and the dog drawing the bear right back to the human. One dog picked a fight with a beaver and wound up needing to be evacuated by helicopter. Another dog ran himself off a cliff. A tower dog had been so aggressive, it wouldn't allow the helicopter with firefighters aboard to land on the helipad. What if Camille turned out to be the dog from hell, bringing terror to the solace I longed for at the tower? I returned her to the shelter and mulled over my options.

Then, five days before my departure, I received a call from the detection aide, a woman who helped to coordinate fire tower logistics in the Peace Country.

"My mom has one too many dogs on the farm," she said. "One of the dogs, Holly, isn't getting along with the other female dogs. You're welcome to take her for the summer."

It felt like an arranged marriage—I needed a temporary dog and she needed a temporary owner.

I drove an hour north to their farm outside Manning. When I turned down their long gravel driveway, three dogs bounded towards my car: a terrifying German shepherd with an aggressive bark and flashing teeth; a huge black monster of a dog that leapt at my driver's window; and a smaller dog, the fastest and most agile of the three. She didn't bark but raced ahead to the farmhouse. I parked, and the dogs circled my car like a wolf pack. The owner called them off, and I got out to greet her. The smallest dog melted into my side, proffering kisses. She was pretty, with a sleek black overcoat, a white belly, and caramel legs. Her brown eyes were like bowls of chocolate pudding lined with black kohl. Her ear felt like crushed velvet in my palm.

"So, Holly is the German shepherd?" I asked the owner, pointing to the aggressive dog that paced back and forth, head low, eyeing me suspiciously.

"No, that's Holly."

I looked down at the adoring, mute dog and she looked up at me. I wondered, this is the dog that's going to save my ass from getting mauled by a bear, or a wild cat? My gut sounded off the alarm, *wrong dog, wrong dog*. But it was too late. The owner leaned down and whispered into her ear. "Take care of her out there, okay?"

"I'll bring her back in the fall," I told the owner, and she nodded.

Holly bounded into the back seat of my car without hesitation, her tongue hanging out of her mouth in a kind of goofy grin. As we drove back to Peace River, I looked back at her every few minutes in the rear-view mirror. Did I trust her to defend me against predators?

That evening, I googled the meaning of her name.

hol-ly
noun
a widely distributed shrub, typically having prickly dark green leaves, small white flowers, and red berries. There are several deciduous species of holly but the evergreen hollies are more typical and familiar.

"When you are feeling the walls cave in on you, call upon the holly," I read from a website on Celtic symbols and meanings. The ancient Celts would plant the thorny crop around their homes and the bright-red berries it bore were believed to bring good luck—they were symbols of vigilance and protection.

"Are you lucky, girl?" I asked her.

I scratched behind her velveteen ear and she let out a contented sigh. *Thump-thump-thump.*

Her tail beat out a joyous song.

After a year in Uganda, I met Aisha, a young Ethiopian woman who came to volunteer in Kabale. Aisha was an undergraduate student majoring in political science at an Ivy League school in the States. Her father was an economist with the UN's African Development Bank. As a child, she moved around the African continent, growing up in both Addis Ababa, Ethiopia, and Abidjan, the capital of the Ivory Coast, spending time in North Africa, and graduating from a private high school in Chicago.

Her father had enormous expectations for Aisha: Graduate with honours. Work for the United Nations. Marry a successful Muslim man. She was driven and ambitious, but also curious to stray out of bounds. That curiosity had led her to Kabale, foregoing internship opportunities with the World Bank and World Health Organization to volunteer with a grassroots organization.

Aisha was African-American, but the locals didn't treat her with any more familiarity than they treated me.

"*Omuzungu!*" they hollered at her in the streets, as though they were hollering at me.

"I'm African too!" she yelled back at them.

We became fast friends. She, Akello, and I often hung out after work, listening to music and chatting. She and Akello connected over their love of hip hop and dancing. They traded dance moves back and forth, a certain sway of the hips, drop of the knees, flick of the arms and shoulders. Not much of a dancer, I watched from the sidelines, impressed and intimidated.

One day, Patricia, Aisha, and I sat on a woven mat on the lawn in front of the doctor's house. Patricia was braiding Aisha's loose black curls into tight rows.

"Quit moving," Patricia scolded her.

"Not so tight," Aisha complained.

We spoke about school and work and love and relationships. Aisha knew about my relationship with Akello, as did Patricia.

"What do you think your parents would say about you dating Akello?" Aisha asked me.

"They know," I said. I'd told them over a Skype call. "They support my decision."

She studied me. I felt self-conscious under her hard, self-possessed gaze.

"You're the freest person I've ever met," she said. "You follow your heart. You take risks. I could never do that. Up and leave a good job. Fall in love with whomever. My parents have serious expectations for me. If I brought home a man who wasn't a Muslim, my dad would probably disown me."

I considered her words. Was it really about freedom? My white, middle-class Canadian upbringing, my parents' support, access to good employment, and a financial safety net protected me in many respects. I could afford to take risks in work, geography, and, yes, even in love. Akello was anything but the boy next door. He was a community college student, a soccer player, and the grandson of poor farmers. I loved him fiercely.

"Don't you worry about bringing him to Canada, though?" Aisha asked me.

Patricia glanced up curiously, though her fingers continued to pull and weave strands of Aisha's hair.

"My friend met a guy in Kenya," explained Aisha. "They got engaged and he immigrated to the States. But he became a totally different person when he arrived. He ran off with somebody else."

I was already aware of the stereotypes, what people must think— Ugandans, Canadians, Americans: that Akello was using me for my money, my passport, my address in North America.

Or, conversely, that I, as a lonely white woman, was using *him* for companionship, comfort, and a kind of "exotic" love. I hated these stereotypes that diminished our relationship. I'd worked in the international development sector for nearly a decade. Many of my friends in Edmonton had met their significant others while working abroad and lived in cross-cultural relationships and marriages. Some of them now had young children. A cross-cultural relationship seemed to bring additional challenges, but Akello and I communicated more effectively than I had with many of my previous Canadian partners. I trusted him.

"We'll make it work," I told her. He was my best friend.

Aisha nodded, but I felt her skepticism. I pushed aside my own worries. Wasn't that part of loving someone? I asked myself. Doubt was natural in any relationship. Love required a leap of faith.

I noticed several hawks sailing in a circular pattern above a few nearby buildings. They rode the hot-air currents to rise up, then kited down towards the earth, then soared up again towards the clouds. I watched them rise and tumble over and over again.

"Look at how beautiful they are," I mused, pointing skyward. "They're playing in the wind."

Aisha, Patricia, and I craned our eyes towards the sky, a pure blue scrubbed free of clouds. We watched the hawks circling silently, writing poetry against the blue.

So much of the way we see and make sense of the world depends on who we are and where we come from. Was I seeing the truth about my life—my relationship with Akello—or a romanticized story that I wanted to believe to be true? Perhaps a part of me didn't want to acknowledge our differences, and the difficulties that lay ahead in our journey as a couple.

"They are not playing," Patricia said quietly, her fingers flying. "They are hungry."

After the training in Hinton, I drove north back to Peace River for a regional orientation that all seasonal staff in the district were expected to attend. I walked into the conference room at the government building in Peace River and felt dozens of eyes on me. Even though the number of women working in forestry had grown substantially over the years, here we were, a handful of women amongst fifty men. I felt, suddenly, acutely conscious of my gender. In my previous career, the non-profit sector, women had hugely outnumbered men. It wasn't a subtle shift for me. My cheeks flushed, I avoided eye contact, and I nervously slid into a chair at the back of the room.

We were lookouts, firefighters, camp bosses, radio dispatchers, and administrative staff. The firefighters were almost exclusively male—I noticed only two women in the Nomex uniform—although half of the lookouts were women. There were fourteen towers in total in the Peace River district. Seven towers located in close proximity to communities, where the risk of wildfire was greater, had already opened in early April. The remaining lookouts would be going to remote fly-in towers where there were more trees than human infrastructure. Our seasons would be shorter, running from May to the end of August, whereas the settlement towers were open from April till the end of September. My colleagues were a mix of men and women, mostly in their fifties and sixties, committed to the lookout lifestyle and the task of watching for smoke. Everyone seemed excited about the season ahead, eager to get settled at their home-away-from-homes for another year on the watch. The lookouts were kind and welcoming, but didn't ask many questions of me beyond "What happened to Alex?" I shook my head, confused, then realized they must've been referring to the lookout whom I'd be replacing this season.

"He was good at spotting smokes," said one of the women. She seemed disappointed Alex wouldn't be returning to the tower.

"How was your season last year?" I asked one of the lookouts.

"A bit tiring, actually," she said, noting that 2015 had been an exceptionally busy season, with higher-than-average numbers of wildfires. I'd later learn that this lookout's previous neighbour, a rookie, had shrugged off his daily tower climbs and failed to report a lightning-caused smoke that grew into an enormous fire that took crews all summer to extinguish. He didn't last the season. Not everyone was emotionally or psychologically cut out for the lookout life, and lookouts wanted reliable and self-sufficient neighbours they could count on. Trust and friendship amongst colleagues was earned slowly, over an accumulation of seasons. I'd have to prove myself out there on the watch.

Maybe I made a good impression on this lookout, though, because she kindly scribbled down some tips for me on a small piece of paper, together with additional items to pack:

Transistor radio with AM frequency—a static crackle over the
radio means lightning
Hot water bottle for cold days up in the tower

Only a week later, she'd become one of the voices on the radio that I admired. Her fifteen-plus years on towers had made her into a calm, confident lookout. I'd hear her expertise in the measured way she spoke over the radio, directing firefighting crews to find hard-to-see, low-lying smokes.

I was one of only two rookie lookouts that season. My colleagues had worked anywhere from four to thirty consecutive seasons on the towers. They were an eclectic bunch: An arborist in his twenties who brought his collection of samurai swords to the tower. A farm kid in his fourth year of engineering who played the guitar and disc golf. A fellow who worked as a professional dog walker in the off-season. One woman who could shoot and skin a moose and said that the fire season was her "social season," and another with a warm, friendly demeanour who called herself the "happy lookout." I felt like a new kid trying to get into the club, trying to get the other kids' approval. I wasn't sure if they'd accept me. Even so, I was relieved to put faces to the names of some of my neighbours. Soon their names would disappear from my tongue, however, as I'd only call them over the radio by their call signs— XMB99, XMD84, XMD76, XMA732, and XMA791.

Me, I would become XMA567.

I cast a side glance at some of the firefighters, yellow giants slouched in their chairs, legs splayed wide. They were ruggedly good-looking: beards, grizzled faces, long, sun-bleached hair. Some were young, probably only eighteen or nineteen years old. Most were in their mid-twenties. Some—the leaders, I suspected—appeared older. They had a more laid-back, seasoned way about them and moved with confidence. Their yellow uniforms were stained black with soot. I avoided their eyes.

During a coffee break, a tall firefighter, probably in his mid-twenties, wearing black-rimmed glasses, approached me and a few other lookouts.

He wore his sun-bleached hair back in a ponytail. His face was tanned, suggesting he'd spent the off-season in a warmer, tropical locale.

"It seems like a really interesting job," he said, referring to the fire tower. "I can see my parents doing it when they've retired. I don't know if I could do it, though. That's a long time to be alone."

His name was Jay. Originally from southern Ontario, he had worked in Alberta as a firefighter for several seasons. This year he was a sub-leader on one of the unit crews, a larger, twenty-person crew. He smiled, revealing a slight gap between his two front teeth. He was handsome and he knew it. Was he looking at me? I looked away. I wasn't sure what to say to him.

"I'm bringing worms," I said awkwardly. "I mean, composting worms."

He nodded as though he was genuinely interested.

"I want to keep a garden at the tower," I said quickly, explaining myself. "Build up the soil."

He looked closely at me when I spoke. I felt something intensely familiar in his eyes. Something kind and gentle. I excused myself to go to the bathroom, though I didn't have to. I sat down on the toilet, feeling flattered, then flustered. And finally, deeply ashamed. Why hadn't I told him: I have a fiancé. I'm working this job to raise enough money to bring him to Canada. My life isn't here. This is not my real life. My life is in Uganda. Why couldn't I speak aloud my truth? I was hiding something. From him, my fiancé—and myself.

One evening during the summer of 2015, a few months after I'd flown back from Uganda to Canada, I ran into an old friend at a bar. An acquaintance, really. He was an artist. An older man. Someone I'd always felt drawn to for reasons I couldn't name. We talked and drank until our friends had left. He told me about the grief he felt over the recent loss of his father, and I told him about Akello and the challenges of immigration, my worry about moving the process forward only to be disappointed by bureaucratic red tape. It felt good to open up to an acquaintance, someone I didn't really know and who didn't really know me, about my fears.

"That sounds impossibly hard," he said.

I nodded blankly and drowned myself in that word—*impossible.*

We sat at the bar, elbows grazing, something dangerous passing between us. We tilted amber pints down our throats until the bar closed. Then we walked back to my apartment and I invited him upstairs. He put on a 1950s love song and we slow-danced in my living room.

I kissed him hard. His kiss, too, was like the booze. Intoxicating. I forgot everything.

I woke up the next morning as reality slipped through the blinds. I felt his foreign body behind mine, and every muscle tensed. The sudden remembering of the blurry events of the night before, my infidelity, hit like hurled stones. I got up and rushed to the bathroom. I knelt down, as though to pray, and stuck my fingers down my throat to force the shame—*I am a bad woman*—out of my body.

Several weeks later, I broke down in a Skype call from Edmonton to Kabale and admitted to Akello that I wasn't ready to get married.

"I can't do it," I sobbed to him like a child.

"Why not?" he asked me. "Trina, what is this really about?"

I couldn't tell him I had slept with another man. I couldn't bear the idea of hurting him, his family, and our community with the truth. Mostly, I couldn't bear the idea of losing him and the shared future we'd envisioned.

"Come home," he said gently. "And we can sort everything out together."

A month later, I flew back to Uganda. When I saw Akello waiting outside the airport, I ran into his long, tree-branch arms. Akello wanted to get married right away, but I refused. We compromised: we'd have a traditional introduction ceremony, a prelude to a wedding in Uganda, to introduce our families to one another. Akello's uncle would host the event in his backyard.

"I don't want it to be a huge event," I stressed to Akello. "I want just a small gathering."

"Of course," he said.

True to Ugandan tradition, three hundred people attended. Men came dressed in suit jackets and ties, and women drew inspiration from a mixture of modern and traditional styles. Some wore cocktail dresses while others wore the *gomesi*, a fancy, slouching dress with pointed, puffy shoulders, which was secured by a large cloth belt.

I allowed myself to be wrapped in yards of turquoise silk. Akello's aunt spun me around like a doll, swathing my waist in the shimmering, floor-length cloth. She draped the cloth over my shoulder, then tied a string of blue beads, a kind of African tiara, around my forehead.

"My daughter!" she cooed, clasping her palms together.

I felt an ocean of eyes on me. My parents and two dozen family friends sat under a white canopy tent. Akello's family, 250 of them, sat under separate tents.

Akello was smartly dressed in a black-and-white zebra-patterned traditional shirt and had woven dreads into his coiled hair. He looked so handsome, so happy. My heart pulsed with love for him. I slipped my small hand into his large one and we danced down the aisle together, shaking our hips, stepping in beat to a punchy African pop song by Judith Babirye, a beloved Ugandan musician.

Family and friends made speeches to celebrate our friendship and the coming together of our two families. They lit fireworks above a table stacked high with ten ginger-flavoured cakes layered with rock-hard orange icing. I knelt on the ground, according to custom, and fed him a piece of cake as he sat above me in a chair. Who was this woman? I wondered. I could hardly recognize myself in that moment, kneeling passively in front of Akello, feeding a man who could easily feed himself. People cooed at me and snapped photographs and I wanted, for a brief moment, to disappear. But then Akello broke Ugandan tradition, knelt his tall body down on the ground, and fed me a piece of cake, which prompted a ripple of startled laughter in the crowd. I smiled gratefully. He was strong that way, always acting according to the culture of his heart. I could only imagine how his Ugandan male friends would later tease him, but he'd knelt down for me anyway.

Plastic glasses of orange soda pop were poured and passed around and a toast was made, and then the music started up and everyone was dancing and for a while I forgot about my fear, and the pain of remembering how I'd betrayed him.

We danced, buoyed by the music, the air of jubilation, and a sea of people dancing around us.

I stayed in Uganda for another two months and we behaved like honeymooners until my departure. I didn't want to leave. I was determined to make our relationship work and be the partner he deserved. I would be a good woman.

PART TWO

PRE-SMOKE

Time was thick, perspectives confined to the middle distance. I felt panic. I needed a horizon. I climbed back up to the perch above the treetops.

—ELLEN MELOY, *The Anthropology of Turquoise: Meditations on Landscape, Art, and Spirit*

CHAPTER FIVE

The ranger backed his truck up to the helicopter. The material contents of my life for the next four months were stacked high in the back of the truck: boxes of food, jugs of drinking water, clothes, bedding, two bins of books, a carving knife, acrylic paints, a yoga mat, a ukulele, and a 12-gauge shotgun.

"Here we go," the ranger exclaimed with a wide grin. His name was Jim. It was too perfect, I thought. Ranger Jim. He was an East Coaster who'd worked in the North for decades.

"Let's go say hello to the pilot," said Jim.

I jumped down from the truck and followed him over to the helicopter, a huge, glossy machine painted Camaro red. It was a medium-sized helicopter with ten seats in the back and two in the front, powerful enough to carry over two thousand pounds. Unit crews, who were considered to be the giants of firefighters, typically flew eight-man crews in mediums. Helitack crews, generally the first responders to wildfires, often flew with four-man crews in smaller helicopters. I was grateful we were taking the larger machine—I worried my gear wouldn't fit.

"Uh, is that your dog?" asked the pilot, pointing behind me.

I looked back over my shoulder to see Holly parading down the runway, her leash trailing behind her. She must have jumped out of Jim's truck window. Oh shit, I thought. "Don't piss off the pilots," they told us at the tower training. Great start, I thought, chasing after her. The vet had recommended giving Holly a mild sedative an hour before the helicopter ride. I kept waiting for it to kick in, for her eyes to close, her head to droop low, but here she was now, hyped up and prancing down the runway with the energy of a puppy. She must've picked up on my frenzied energy. AWOooooo! she yowled comically. I apologized, but the pilot just laughed.

"Let's get this show on the road!" he said playfully, motioning for us to start hauling everything in Jim's truck over to the helicopter.

The pilot expertly stacked the boxes in the back of the machine. Twelve hundred and eighty-five pounds—that's how much my material life weighed. Jim scribbled passenger names and the total weight onto the flight manifest. The pilot bent down and heaved Holly up into his arms, nudging her into the dog kennel in the back seat. I saw the flash of fear in her eyes.

"Take the front seat," Ranger Jim instructed me. "That way you can get a closer look at what's around your tower." I pulled myself up into the front and my hands fumbled with the seat belt as the pilot started up the machine. The rotor blade began whirring. The noise thundered, then deafened, so I reached for the headset above my seat.

My heart hammered in my chest and throat and ears. The doors closed and the blades accelerated. The pilot grinned and flashed me the thumbs up. I gave him the A-okay sign, though I thought for a second about drawing a finger across my neck and putting a swift end to everything.

We peeled off the earth. The helicopter hovered for a split second above the ground before ascending straight up, up, up, up. Adrenalin coursed through my body like the mountain streams after winter's thaw, full-bodied, eager to move, to go somewhere, anywhere. I gawked at the earth below, the dried grass blowing like a wild mane of hair. Civilization shrinking to childlike proportions: miniature airplanes, buildings, cars, and highway. From a bird's-eye view, how insignificant the civilized world seemed. Everything slid away.

Goodbye, farmhouse. Goodbye, power lines. Goodbye, highway.

"26, this is Tango Whiskey Victor," said Ranger Jim, transmitting a flight message to the radio dispatchers in Peace River.

"Tango Whiskey Victor, go ahead for 26," replied the dispatcher.

"You can check TWV airborne off the Peace River airport," said Ranger Jim. "We're heading north with myself and passengers Trina Moyles and Holly the Tower Dog on board."

"That's all copied, 26."

Leaving. It wasn't hard for me to do. I was reminded of my nineteen-year-old self, boarding an airplane to Central America for the first time, and my twenty-seven-year-old self, sleeping overnight in Heathrow Airport on New Year's Eve and boarding a half-empty jumbo jet for Entebbe, Uganda. Only now I was flying towards a destination that didn't exist on any map. No man's land. Few had journeyed to the fire tower. It was a place that barely registered in people's imaginations, as distant and intangible as Antarctica.

The boreal forest unfurled below like a handwoven rug, rolling into the blue band of horizon. I had flown above far-flung landscapes before—rainforest and desert and mountains and oceans. But I'd never flown north of the fifty-sixth parallel, the place where I grew up. I'd never seen the boreal forest from the perspective of a soaring bird.

The tapestry blended together aspen, birch, pine, and white and black spruce. The deciduous trees hadn't yet donned their leaves, and the birch buds blotted the landscape with a meek red hue, barely noticeable. Sunlight illuminated the stands of pine, which appeared more golden than green. The black spruce was snarled and misshapen, ugly as a snaggle tooth. Up north, beauty is a crooked thing, like a bone that's been broken again and again.

We flew north of the Notikewin River, a tightly coiled, snaking waterway that takes days to paddle. I peered below and spied a moose, bedded down in the grey brush. Then a pair of trumpeter swans, two white specs afloat on a brown, murky lake. A small black dot moved along a seismic line, a narrow corridor used to transport and deploy geophysical survey equipment: *Ursus americanus*, an American black bear, tiny, minuscule from above. I hoped the seismic line wouldn't lead to my tower.

The boreal wasn't pristine—it wasn't as untouched as I had dreamt it would be. Human influence was evident in the number of straight lines savagely slicing up the forest: cutlines, seismic lines, winter roads, access roads, all running parallel and criss-crossing. It was scissor work. Hardly any plant life grew from where machines had scraped down to mineral soil. It would take decades for many of these cutlines to grow life again. Many of the roads led to square clearings where men had built machines to penetrate deep into the soil and suck up what lay beneath. Crude oil and sour gas. Many of the well sites had long been abandoned. Larger companies sold off depleting fossil fuel reserves to smaller companies, and small companies often couldn't foot the costs of environmental reclamation.

Pathways like the seismic line favoured some wild things: wolves, cougars, and bears, predators that could zip easily along a line while on the hunt. But they were death traps for woodland caribou, who were hunted by the wolves with increasing ease and frequency. That and the logging industry had severely fractured and depleted their habitat. Caribou dwelled in old-growth spruce and fed on intricate tufts of lichen that grew and dangled from the trees' lowest, oldest limbs. It took decades to grow old coniferous forest and lichen and ensure the survival of one of the boreal's oldest ungulates.

"Will I see any caribou up at my fire tower?" I'd asked my dad before leaving.

"Not likely." He shook his head solemnly.

According to my father, who'd flown over caribou herds in the Peace Country for three decades, the majority of herds would soon be reduced to the symbol on the tails side of a Canadian quarter. There was a good reason biologists called them the "grey ghosts" of the forest.

"There it is," said the pilot, pointing towards a gentle, sloping hilltop.

I strained to see, but the vibration of the helicopter dislodged my vision. Then everything came into focus: the fire tower, a thumbtack half pressed into the rolling carpet of trees. We drew closer and closer. I felt panic rising in my chest.

"So, Trina, you'll notice that you've got a lot of coniferous around your tower," crackled Ranger Jim's voice over the radio. I glanced over my shoulder and the pilot gestured out the window.

But no, I hadn't noticed. I couldn't take my eyes off the tower. I could barely breathe as my gaze locked on that silver filament rising up out of the earth. H-O-M-E.

I didn't yet know that old-growth coniferous forest was akin to a kerosene-soaked woodpile, their low-lying limbs and resin-drenched needles designed by nature to burn. Ranger Jim was warning me: stay alert, because a close-range fire would burn hot and fast and threaten to burn over my tower. The lookout to my north had been air-evacuated when the wildfire came within kilometres of his tower in the summer of 2015. It had been a record-breaking year for wildfires in Alberta.

But I didn't know that yet. I didn't know anything, really, about what I was getting myself into.

We rode up a ridge and the forest split open below, finally revealing our destination. The red-and-white cupola at the top of the steel tower was a beacon against the palette of grey-and-green bush. A yellow cabin sat beneath the tower, its windows boarded up. A smaller building, an engine shed, and a large solar panel were adjacent to the cabin. A long rectangle of forest had been cleared, several hundred metres away, to serve as an airstrip for helicopters and small airplanes, and a trail through spruce connected the cabin to the airstrip. The clearing, the cabin, the airstrip, the tower—it wasn't majestic. It was ugly in fact, an eyesore that I wanted to disappear. Wrong tower, I wanted to tell the pilot.

"26, this is TWV," said Ranger Jim to the dispatch.

The helicopter descended way too fast. I sucked in a sharp breath.

"We're final for the fire tower."

———

I looked around at my clearing in the woods, the bleached-blond grass surrounding my cabin. Wind whipped my long hair about my face as I heaved the heavy food bins to the cabin door.

Ranger Jim was in a hurry.

We entered the small yellow cabin, not bothering to take off our boots at the front door. It didn't matter—the place was filthy. A set of muddy footprints tracked across the linoleum floor. Ranger Jim sat down at the round wooden kitchen table, which was covered with a cheap plastic cloth the colour of Astroturf. He balanced a pair of reading glasses on the crook of his nose as he checked off nearly seven pages' worth of items on the to-do list.

✓　Assemble the Osborne Fire Finder in the cupola

Ranger Jim had awkwardly struggled into one of my extra-small climbing harnesses, looking uncomfortable with the maxed-out straps cinched up, but he climbed anyway, his reach spanning several ladder rungs. I followed him up with the spare harness, cursing myself at every rung, fear pounding in my ears. At sixty feet, I was finally above the forest. Five minutes later, I reached the top.

The cupola was much smaller than I remembered from the one at training. Perched on a wooden stand, the large Osborne Fire Finder ate up most of the space in the tiny dome. A cross-shot map had been tacked up to a bulletin board on one of the eight walls. I would be the first lookout to use the newly built cupola, so, except for the map, the walls were bare, the only furniture an old padded swivel office chair. I'd have to scout out the forest and make note of important landmarks, their bearings and distances, so I could place myself on the map.

It was a tight squeeze up in the sky. But the view! The forest seemed to go on forever in every direction. I felt very small in the tower, a speck of dust, a single seed, defying gravity.

"Which way is north?" Ranger Jim asked.

Tired and disoriented from the climb, or perhaps from the suddenness of dropping down out of the sky, I pointed south. He shook his head. I was off to a bad start. What must he think? They'd hired a lookout who couldn't tell north from south.

I followed Ranger Jim back down to the earth and the work continued.

- ✓ Open cabin
- ✓ Turn on propane and gas appliances
- ✓ Test communications equipment

Outside, I'd unhinged the large wooden shutters covering the cabin's windows. Fine, papery wings fluttered out, cascading to the ground like confetti. Hundreds of dead or barely alive butterflies. The wings belonged to *Nymphalis vaualbum*, or Compton tortoiseshell, a species commonly found in Canada's northern boreal forest. Flung wide open, they were the colour of a forest on fire. Closed, they were mottled tree bark, hands folded in prayer. I knelt to pick up one of the tortoiseshells and burnt-orange dust stained my fingertips. The butterfly slowly opened her wings. It was no small feat, I thought, such a fragile creature surviving the harshest season in the boreal.

- ✓ Set up rain barrels

Ranger Jim and the pilot struggled to push the large, empty plastic rain barrels through the shed door. They propped up a 350-gallon barrel on a wooden stand beneath the cabin's eavestroughs. The pilot emerged from the shed with an archaic aluminum bathtub. "A bathtub!" he said, grinning boyishly. He helped me rig up another eavestrough using lost-and-found items in the shed to catch rainwater off the front porch roof, and we placed the old tub beneath. I wanted to grow my own food, so I'd need as much rainwater as possible. Every barrel echoed hollow. Until the rain fell, I'd have to survive on what we'd flown in: five 18-litre jugs of drinking water and another twenty-five gallons of water for cleaning, washing dishes and clothes, and bathing. My body already felt filthy with the knowledge that I wouldn't be able to justify a proper hot shower until the skies thundered and darkened, let moisture loose, and filled up my rain barrels.

Holly dashed excitedly to and fro. She prowled the perimeter of the clearing, nose low to the ground, chasing invisible scent trails. What wild things had lurked here before us? Moose? Bears? Ranger Jim pointed fifty metres downslope, towards a stand of naked willows and black spruce.

"A few years ago, I was out delivering groceries and we watched three cougars walk along the edge of the bush. They didn't seem to mind us. They just walked on by. Right there."

The hairs on my arms rose up, sensing their ghostly cat figures, but I kept my fear to myself.

Holly ran laps around us. She wiggled her hips back and forth and let out multiple victory howls—Awoooooooo!—and scampered in and out of the cabin. She seemed right at home.

I wished I felt the same, but I didn't. They'd replaced the steel fire tower over the winter, trucking in equipment and parts when the muskeg ground was frozen. The yard was a mess of frozen tire ruts, exposed earth, and bits of steel and broken pieces of the old cupola. There were cigarette butts strewn about on the yellowed grass by the front porch. And the garden beds had gone feral, overrun by the roots of prickly wild rose bushes. I had expected a quaint and cozy, picturesque cabin in unspoiled forest, not a ramshackle home atop a pile of mud and thorns.

The cabin felt as though it had been occupied by a hoarder. Odds and ends—tools, scrapbooking supplies, bakeware, old batteries—occupied every shelf, cupboard, and drawer. Someone had hung hand-sewn ruffled half curtains, buttercup yellow with orange-blossom flowers, over the windows. A single book sat on the shelf in the bedroom: *The Snow Walker* by Farley Mowat. A quote by Groucho Marx had been tacked up by the radio desk: "Outside of a dog, a book is a man's best friend. Inside of a dog it's too dark to read." Strange artwork hung on the walls. Someone had used Photoshop to create an overlay image of a forest and a tropical island with white sand beaches. A longing to be somewhere else, perhaps.

Ranger Jim shuffled his papers and tucked the reading glasses into his shirt pocket.

"Well, we best get going. You'll be all right?" he asked.

I nodded meekly, swallowing down the feeling of inadequacy, and watched the six-foot-tall ranger go. The pilot was already in the machine, the blades firing up. Ranger Jim turned and offered up a hand as a final farewell.

"Good luck, little one!" he hollered above the sound of the machine.

A laugh spilt from my lips. I wanted to finish his sentence for him, acknowledge what he wasn't saying. "Good luck and don't get eaten by a bear!" or "Good luck and see you in a week when you're crying for us to take you out!" He'd meant it, no doubt, in a playful way. Not unlike the way my Cuban friends called me Trinita, the *-ita* suffix a term of endearment. Trinita. Little Trina. *Little one.*

I felt insignificant, unprepared, standing alone on the front steps of the cabin. The helicopter blades churned then thundered. I looked down at Holly and she craned her head to look back up at me, her black-and-white-tipped husky tail beating out a song on the step. *Thump-thump-thump.*

We're in this together, Human, she seemed to say.

When I looked up again, the helicopter was hovering above the ground, the dead grass a wild mane whipping around, the rubbery black spruce bending back and forth. The pilot spun the machine, pointing south, and away they went, flying above the forest, shrinking to the size of a dragonfly.

I flung my body, arms outstretched, into the void.

"GOODBYE!" I screamed.

Holly leapt and gave chase, playfully nipping at my legs, yowling madly, dancing at my feet.

I spun around, feeling the enormity of the forest, the space, the distance between civilization and wilderness, the known and the unknown. I felt vulnerable and more powerful than I'd ever felt before. We don't have an adequate word to describe what was charged and running feral through my veins: a blend of fear and freedom. I'd never felt so terrified. And yet never so alive.

"I'm excited about the silence," I had previously said to my friends and family and my love fourteen thousand kilometres away. But now I held my ear up to the forest and discovered the truth—that silence did not exist. That I had not stepped into a vacuum of sound, that the forest was full of noises, foreign ones. The naked aspen creaked and shook. A hollow drumming sound came from deep in the woods, the sound of a male ruffed grouse beating his feathers against his chest—a love song of sorts. I heard a howl too close for comfort. My heart picked up its tempo. No—it was only the sound of the wind playing the steel guy wires of the tower like a violin. But nothing was louder than my own thoughts.

Don't be weak.

Don't let the forest burn down.

Don't let them down.

Don't let your love down.

Who do you think you are?

CHAPTER SIX

I woke up at 6 a.m., confused, as though I'd already forgotten yester-day's flight over the forest. My body had arrived, but my mind hadn't yet caught up.

I got out of bed and opened the front door, stood on the front step, and felt the isolation come rushing in at me. Reality sank in. You are here. You are here for the next four months, without break. And you are very, very alone. The thought was a run-on sentence that played over and over in my mind:

you are alone you are alone

I grabbed a canister of bear spray and walked slowly to the out-house, twenty metres away from the cabin, glancing from side to side, scanning the edges of the bush for a bear, a cougar, a wolf. I was terrified.

Holly, on the other hand, trotted about, zigzagging back and forth, nose to the ground. She was her usual relaxed, jovial self.

I was relieved when the cabin door shut behind me, as though the poorly insulated walls were enough to buffer me against the wild. I replaced the green plastic table covering with a red wine–coloured cloth that I'd brought back with me from Uganda and imagined drinking coffee with Akello at our large wooden table in Kabale. I glanced at the clock and did the math: Akello was ten hours ahead. He'd probably be harvesting vegetables from our garden right now, or maybe simmering a soup over the gas stove. I longed to hear his voice. I grabbed my laptop and called his cellphone over Skype.

"The person you are calling is unavailable," said a woman's voice.

My heart sank.

I cranked a hand grinder to pulverize coffee beans, boiled water for oatmeal, unpacked a few bins, and used a damp rag to dust the countertops. The cabin was a mess, but I wouldn't have time to clean today. My morning weather report was due in an hour. I had to pack a bag with food and water and climb before 9 a.m.

Everyone in the forest—including lookouts, firefighters, pilots, and wildfire rangers—was on high to extreme hazard alert. Midday temperatures had been climbing into the 20-degree-Celsius range, soaring generously above the average temperature for late April.

The phone rang unexpectedly. Who was calling so early?

"Hello?" I answered nervously, almost suspiciously.

"Uh, hello there," said a male voice, gravelly and cheerful. "Welcome to the forest. I'm your neighbour to the southeast. Just wanted to welcome you to the neighbourhood."

The man's name was Ralph, although one of the managers in Hinton had referred to him as the Grandfather of the Forest. He was known by many lookouts, rangers, and firefighters in Alberta. The eighty-year-old had worked on fire towers for three decades, and well before that he'd begun his career as a firefighter in the early 1960s. He knew the forests like the back of his hand. Come autumn, he and his wife, Bea, hosted groups of American and European hunters out in the bush, guiding them towards moose, elk, deer, and black bears. Ralph had been trained by Sam Fomuk, a legendary lookout in Alberta's fire tower history, who worked nearly fifty years scouting for smokes. I'd never forget the stories I'd come to hear about Sam Fomuk: A bolt of lightning once blew his cabin radio to bits and he'd had to rewire it by hand. Sam often went out with only a couple of bags of flour and rice and a fishing rod—and he lasted six, sometimes seven months in the bush, chartering his own helicopter in November, long after the fire season had wound down, to get picked up. One ranger remembered that Sam would shave his head before going in and come out with his hair looking like a "dandelion gone to seed."

Ralph was unique because he belonged to a generation of lookouts with one foot in the past and one foot in the present, having worked long summers where his only sources of connection with the outside world were the two-way radio and handwritten letters, and now having access to cellphones and high-speed Internet. He'd seen many changes over the years, but the task remained the same: watch for smokes and "catch 'em small."

Some things about working at a fire tower, I would learn, would never change. Like the fear you feel on the first day on the job. It seemed as though Ralph could sense my anxiety from over fifty kilometres away. The phone call felt like a good omen for my first day at the lookout—an unexpected kindness.

"Call me if you need a hand," he said gently.

I promised that I would.

"Cross-over conditions expected," said the radio dispatcher for the a.m. weather forecast. *Cross-over.* Those three syllables sounded the alarm for everyone in the forest. Cross-over occurs when the relative humidity drops below the temperature. The combination of low humidity and high temperatures signals the possibility of disaster to wildfire managers, increasing the chance of fire ignition, "multi-starts" (multiple fires igniting at the same time), and the rate of spread.

"The forest is so dry that if someone so much as lets one rip—oh, Christ, we're *screwed*," I'd overheard a ranger say casually at the orientation.

Looking out over the matchstick forest, I sensed that my first few days on the job wouldn't be uneventful.

In fact, I could already see the evidence of one fire burning out of control. A dark stain loomed on the eastern horizon, gargantuan smoke columns produced from a firestorm—multiple fires that had burned together to form one massive fire complex—that was raging hundreds of kilometres away in Fort McMurray, Alberta. Only days ago, it had started in the Horse River wilderness area, southwest of Fort McMurray. The fire was thought to have been accidentally ignited on May 1 by recreationalists riding an ATV, the machine's hot engine sparking flames in tall grass. Within hours, the Horse River Fire was burning beyond control, propelled by sixty-kilometre-an-hour winds. This was the kind of fire that firefighters call a "fuck-off fire" because they have no choice— it's too hot, too volatile, and they have to stand back, get out of the way, and wait for the air tankers to drop loads of chemical fire retardant to slow the spread. Much to the frustration of some Albertan wildfire managers and firefighters, the Horse River Fire would become known as the "Beast," a sensationalized name that quickly caught the attention of the media and brought to mind an evil, animal force terrorizing the city. But the Horse River Fire wasn't inherently evil. It wasn't a beast. It was a large-scale, campaign wildfire—not unlike many dozens of fires that had burned before it—that just so happened to ignite and spread in the way of a city.

There was no slowing down the Horse River Fire. On the very day it ignited, the mayor of Fort McMurray began evacuating neighbourhoods. Today, May 3, the entire city was being siphoned out: ninety thousand people caravanning north to oil camps or south to neighbouring towns and cities, where makeshift refugee camps had been set up.

My supervisor had sent all of the lookouts an email about the crisis unfolding in Fort McMurray, but I wasn't seeing what my friends and the public were seeing. I had deleted my social media apps for the summer and wasn't reading news articles online. The Horse River Fire, for me, was only a distant bruise on the horizon. It wouldn't be until much later, when someone delivered a newspaper to me, that I came face to face with the images of bumper-to-bumper traffic, of people fleeing the city as the flames torched trees along the highway. A coral sky on fire looming eerily behind a row of immaculate cookie-cutter homes in suburbia. I would be moved by the story of a young woman who galloped away from the blaze on horseback, leading several of her beloved equine companions to safety.

I examined the bruise from afar, through binoculars, wondering, could it happen here?

It was hard to contemplate such a calamity from a place of relative calm. My view from the tower felt so expansive it stole the breath from my lungs. Songbirds stirred and trilled from the naked tree boughs below. Stands of parched pine gleamed olive and ochre and burnt orange. To my south, I noticed a flock of sandhill cranes circling high in the sky, soaring slowly in a seemingly chaotic swarm. They warbled a song that sounded prehistoric. The migratory birds were flying home to their breeding ponds in the northern boreal and the Arctic tundra beyond. Forty kilometres to my northeast, I gazed into a small Métis farming community where farmers had cleared the forest to plant canola. From the fixed vantage point of my tower, it was hard to see much of the human impact on the land, although I knew many of the trees in northern Alberta had been cut and burned, and the muskeg drained, to make way for canola fields. I could make out several houses that sparkled white like sugar cubes in the sun's rays.

And then my heart kicked up a notch.

I spied several dull, greyish-yellow-coloured clouds floating suspiciously above the farmland. Was it smoke? I watched for several long, agonizing seconds. My hands trembled. Should I call it in? I glanced down at the government-mandated PRE-SMOKE form, which we used to document and report information on potential smokes: bearings, distances, behaviour, and colour. I contemplated recording what I was seeing. But was it really smoke? I remembered a ranger's words from training: "Smoke curls. Dust hangs. Knowing the difference between the two is all in how it moves."

"It's dust," I said aloud to myself, watching the yellowish clouds slowly dissipate. "Just dust," I said again, trying to convince my pulse to slow down. I was afraid to miss a smoke. Afraid to let the forest burn down.

In Alberta, 1,400 wildfires, on average, ignite every year.

In Canada, 10,000 wildfires break out across the country annually.

In the U.S., a staggering average of 100,000 wildfires light up every year, the higher number due largely to the higher population density.

What causes the vast majority of wildfires in North America? The answer is, not surprisingly, people. In Canada, humans cause approximately 50 percent of wildfires. According to the U.S. Forest Service, human activity is the cause of a whopping 85 percent of the nation's wildfires.

Every wildfire has a story.

Hunters, or day hikers, carelessly leave a campfire to smoulder out, but the wind picks up and sends sparks flying into the bush. A brush pile, left unattended, smoulders in a farmer's field and spreads into nearby dry grass. A woman tosses a still-lit cigarette out of the window of her car into the ditch. A father and son enjoy an afternoon of driving their ATV in a wildland recreational park then decide to go off-roading, and the extreme heat from the engine catalyzes a flame in the tall, dry grass. Humans cause most wildfires unintentionally, by pure accident, but there are incidents of pyromaniacs igniting them. The RCMP believes that an arsonist caused the Slave Lake Fire, a wildfire that decimated one-third of the town of Slave Lake, Alberta, forced an evacuation of 7,000 people, and caused over $700 million worth of damages—destroying over 400 homes—in May 2011. No arrest has ever been made.

Wildfires have been triggered by industrial explosions and equipment malfunctions and poorly maintained electrical power lines toppling down into the trees.

Once, a frightened tree planter, standing off with a black bear, shot off a bear banger—essentially a firework—to scare away the creature. He aimed too high, sending the bear banger into the nearby bush and igniting a wildfire that grew to the size of three hectares.

Surprisingly, a fair number of wildfires are caused by wildlife—squirrels, birds, even snakes climb or perch on power lines, get electrocuted, and catch on fire. Years ago, in the Slave Lake district, a black bear cub shimmied up a power pole and was violently zapped. The black ball of fur went up in flames, caught fire in the dry brush, and a wildfire was born.

The causes of some wildfires are unknown, mysteries to be pondered—and rarely solved. But sometimes they have bizarre, even scandalous endings. In 2002, a U.S. forest ranger in Colorado confessed to a court of law that she had accidentally started the Hayman Fire, a 100,000-acre wildfire, by lighting an illegal campfire. Strangely, the ranger had been on patrol in Pike National Forest, enforcing a campfire ban due to the extreme drought conditions. Originally, she claimed to have discovered an unattended campfire that was spreading out of control, but officials found so many inconsistencies in her story that, through further interrogations, the truth slipped out. The ranger, suffering from a broken heart, had stoked the fire to burn an old letter from her estranged husband. The wind kicked up the flames, pushed them into the bush, and they erupted into the second-largest wildfire in Colorado's history, forcing thousands to evacuate their homes. The ranger, an eighteen-year veteran of the U.S. Forest Service, pleaded guilty to arson and lying to investigators and received a federal sentence of six years in prison, along with an additional sentence from the State of Colorado for twelve years in state prison.

There are hundreds of ways to ignite a wildfire. The ingredients are simple: oxygen, heat, and fuel. On a windy day, during peak drought conditions, everything is ready to burn: grain stubble, wild grasses, old pine, brush piles, and even, apparently—love letters.

Climate change has become a major driving factor in fire behaviour. Scientists are predicting an average global temperature increase of one to three degrees Celsius over the next century. The effect of climate change is already widely evident: the loss of glacial ice, rising sea levels, shifting plant and animal habitats, and earlier blossoming of trees and flowers. As for wildfires, scientists predict that they're only poised to get worse. Even in parts of the world where wildfires aren't built into the forest's ecological design, where they don't naturally belong—in rainforests, for example—they are becoming more frequent. Rainforest communities in Nicaragua, Honduras, and Brazil are reporting higher frequency of sudden, destructive wildfires. Climate change has resulted in warmer winters and hotter, drier summers, building the conditions for bigger, hotter, fiercer wildfires in the boreal forest—and around the world.

On my third day at the tower, I called in my first smoke. It emerged from the north, a white smear against an azure sky. Not dust. Definitely not dust. I tried to gauge the distance. It appeared to burn well beyond the farmland, stemming on the far horizon. Probably forty to fifty kilometres away, I estimated. I swivelled around the Fire Finder, peered through the scope, lined up the shot, and scribbled down the bearing. My hands fumbled for the radio mic. My tongue was heavy, my throat already rusty from disuse. I felt a fear akin to the fear when speaking in front of a crowd of people, but much worse, because I couldn't see, or know, who was listening.

"XMA26, this is XMA567 with a pre-smoke," I said nervously into the radio mic, knowing that those two syllables—*pre-smoke*—would set off a multi-faceted response system, dispatching helicopters, firefighters, and air tankers. It took less than half a second for the dispatcher to respond.

"567, ready to copy," she said calmly, the way a kindergarten teacher would answer the raised hand of a five-year-old student. Me, on the other hand, I was vibrating with tension.

"I have a bearing of 30 degrees and 10 minutes," I said, trying to keep my voice from shaking. "Forty-five kilometres away. The smoke is white and coming straight up."

"That's all copied, 26."

The radio exploded with voices. Voices of firefighting crew leaders, pilots, fellow lookouts, and the Bird Dog, the name they gave to the air-attack officer, a ranger who flew above large wildfires and commanded the air show, directing helicopters and air tankers where to drop buckets of water and loads of chemical retardant. The whole gang was heading to my smoke on the horizon. Since the Horse River Fire had taken off a few days ago, Forestry didn't want to take any chances. They'd extinguish anything and everything that remotely resembled a wildfire, even smouldering brush piles in farmers' fields or small campfires.

But this wasn't a campfire. The smoke was huge, even so far away.

"26, this is Alpha Lima Victor," said a male voice over the radio.

"Go ahead."

"Yeah, um, we're not seeing anything at the coordinates you gave us. There's a pretty big smoke coming up from fire in High Level, but it's burning a good eighty kilometres away. I'm thinking that's probably what she saw."

What she saw. She, as in me. My cheeks flushed with shame.

At the command of the duty officer, the firefighting crew landed on my airstrip. I watched one of the firefighters, probably the leader, approach the tower. Fifty metres away from the cabin, he tossed a hand up in greeting.

"Sorry about that!" I yelled down.

And just as soon as I spoke the words, I regretted them. Sorry? Why was I always apologizing?

"No worries," said the crew leader, a burly fellow in his mid-thirties. "Everyone is pretty skittish right now. Better to call 'em if you're not sure."

The problem is, I wanted to tell him, I'm not sure about anything.

When Akello and I moved in together, we rented a spacious two-bedroom apartment set in the sloping hills in Kabale, chosen because of its proximity to my work. I was now acting as the coordinator for groups of U.S. medical and public health students who arrived on a monthly basis. I met daily with the students and often debriefed with them about their practicum rotations during the evenings, so I needed to be close by. Our compound was situated next to the White Horse Inn, one of the oldest, fanciest hotels in town, in an affluent neighbourhood. Rent was 500,000 Ugandan shillings a month, or $200 American, inexpensive by North American standards but an exorbitant amount if you measured it against the average monthly salary of a Ugandan.

Akello had serious reservations about the place.

"*Eh, eh, eh!* It's too expensive," he said.

"I know, but I can afford the rent," I said. "It makes a lot of sense for my work."

Admittedly, I also wanted to live somewhere with familiar comforts: a flushing toilet, hot water, an inside kitchen with a sink and gas stove, a second bedroom for guests, and space to entertain friends and family for meals. I was writing stories, articles, and essays on a regular basis and wanted a space to set up a writing desk to begin my book.

As our lives became more entwined, tensions between Akello and me began to surface more clearly. My own reality as a white, foreign woman afforded me so many advantages—access to a good-paying job with a U.S. organization for one. Meanwhile, Akello, who broke his back at the garage every day, worked odd construction jobs for his uncle and struggled to rake in mere shillings. Even though I was considered low-income by Canadian standards, I had the means to live comfortably in Uganda. The colour of my skin gave me more power in Uganda than Akello, who was born there. The tragic irony wasn't lost on us. I had economic power and choice, whereas Akello had significantly less mobility in the world. It was impossible to deny or ignore how these realities influenced our bond, despite our best efforts.

"My love," he often told me, "even if I have only a shilling to my name, I'll put that money towards our relationship."

And that was true. When money surfaced, he bought groceries, or offered to pay for our meal. But every month, I had to pay for the rent and utilities and the majority of our household expenses, which hurt Akello's pride—even if he didn't say it out loud. For me, being the breadwinner built a sense of oversized responsibility. I began to feel the pressure of our relationship in a new way.

Some of our Ugandan friends, or people who I thought were our friends, were jealous of the move into the affluent neighbourhood. It changed my relationship with several colleagues at the clinic. They spoke unkindly about us, forecasting our doomed relationship.

"It will never work," many said. "That rich *bazungu* lady will leave him, surely."

I only heard about these comments second-hand, through Akello, or Patricia, as often they'd be uttered behind my back, or in fast Rukiga, which I couldn't understand. We settled on the mantra: "Let people talk." We shrugged off people's comments, the gossip and jealousy, as much as we could, and invited our friends and family into our home, sharing tea, coffee, and meals together. But it was hard to shake the feeling that people didn't believe in our relationship, that we were the butt end of people's cruel jokes.

Despite the dynamics we were dealing with, life, for the most part, was very sweet. In the early mornings we woke to the sound of songbirds, hundreds of them, gathering in the mammoth banyan tree across the road. I called the tree *mukaka*, grandmother. Her roots were densely tangled, deep, permanent. Her branches were so large they cancelled out the sky. Akello would slip out of bed and make coffee on the gas stove. He laid out a bunch of bananas, what Ugandans call yellows, and a loaf of bread on our low coffee table.

He was an anomaly: a Ugandan man who loved to cook. Truthfully, Akello was more domestic than I was. He did the lion's share of dusting and mopping the floors, and often sang in the kitchen while chopping vegetables and making soups, curries, and omelettes. He occasionally tolerated rude comments from Ugandan men for doing "women's work."

"You're becoming a woman," they said. Or, worse, "You're turning into a *muzungu*!"

The Ugandan women, on the other hand, praised him for helping with the cooking and cleaning.

"Eh! Trina, you are lucky!" said the woman who lived next door. "You've got a good one!"

When I had a deadline for an article or story, he'd let me work at my desk, clacking away on the computer, insisting on making dinner or doing the laundry by himself. Akello was incredibly supportive of my writing. He saw the fulfillment it brought me, the satisfaction of publishing an article, or the joy I felt while researching and writing *Women Who Dig*. Akello encouraged me to travel to India in 2014 to facilitate research with rural communities about gender and agriculture, and sent loving text messages from afar. He never once made me feel guilty for pursuing my writing, which kept me up late at night, struggling to meet deadlines.

"What's good for you is good for our family," he often said.

One morning, while we lay in bed, bodies intertwined, birdsong filtering in through the open window, the words just fell out of my mouth.

"Let's get married," I said.

He opened his eyes.

Days at the tower were long and taxing. The sun pounded down on the forest. Day after day, the cupola occupancy level was designated extreme. "Extreme" meant I would have to spend eleven hours in the cupola, from 9 a.m. till 8 p.m., with only a brief thirty-minute break to prepare a quick lunch, refill water bottles, and gather temperatures for the afternoon weather report.

I began to feel more robotic than human.

Every day was a replica of the previous one: Wake up at 6:30 a.m. Make coffee and oatmeal. Wash the dishes, sparingly, with a damp cloth. Make my morning climb to check for smoke, visibility, and cloud cover. Do my morning weather report. Pack lunch and water and snacks for the day. Try to call Akello: bad reception or, worse, no answer. Climb the tower for 9 a.m.

Watch for smoke.

12:30 p.m.: Climb down. Go pee. Refill water bottle. Eat lunch. Pat the dog. Take weather calculations. Rush back up the ladder to deliver my weather report at 1 p.m.

Watch for smoke.

Try to read a book, Rebecca Solnit's *A Field Guide to Getting Lost*, but the words become ants swarming the page. By mid-afternoon the cupola is hot. Too hot. Bloody hot. Solar oven hot. Remember watching a demonstration in rural Nicaragua, a woman frying an egg under the glass lid of a solar oven. Consider the glass windows of your cupola.

Think to yourself, I am the egg.

Pee in a yogurt container. Feel ashamed. Fling your shame out the window.

Watch for smoke.

Want to sleep. Poke yourself in the ribs. Stay alert. Watch for smoke.

At 7 p.m., stand by the radio. Listen for your call sign—567—and your occupancy level for the following day: extreme, high, moderate, low-moderate, low. But what was the point? I already knew my fate. Another day without rain meant another eleven-hour day in the cupola.

Watch the vanishing light make the forest shadowed and golden and beautiful.

Climb down to the dancing dog at the bottom of the tower. Feed dog. Feed self. Put on gardening gloves and hack at wild rose bushes in the garden beds. Pull weeds. Build a compost heap. Swat mosquitoes. Haul water in five-gallon pails from a nearby swamp.

Plant seeds.

Retreat to the cabin. Run a damp cloth over exhausted body. No rain, no shower.

Blow out the candle. Listen to the dog, already asleep, her chest rising and falling.

Close eyes. Push away the loneliness.

Dream deep. Dream hard.

Wake up. Do it again.

At one o'clock in the afternoon, the sun was baking the forest at 25 degrees Celsius. The wind whipped the trees back and forth. Cirrostratus, a sheer layer of high-altitude cloud, spilt across the sky like milk on a blue tablecloth. I delivered my afternoon weather report to dispatch and poured myself a cup of coffee, then placed the Thermos back on my Fire Finder, which doubled as a coffee table. The wind howled, the cupola shook, and the coffee in the Thermos sloshed back and forth. My nerves were already shot, but my eyes plodded along the horizon, scouting for smoke.

A wall of ominous yellow-and-grey smoke was advancing from two hundred kilometres away to the southwest. It wasn't a singular smoke, rising up like a ribbon, rather, it was a curtain of smoke smothering the horizon. I became aware of the smoke in the peak heat of the day, as the source, a distant wildfire burning out of control in British Columbia, continued to edge closer. The fire already covered over two thousand hectares and was rapidly spreading into Alberta. One hectare, the measurement used in the world of wildfire, is equivalent to two football fields. I did the math. The fire was four thousand football fields. It was roughly thirty kilometres square, or the equivalent of several small towns stitched together. I couldn't keep my eyes off the smoky substance in the distance. Part of me wanted to fly closer to the wildfire, to witness the flames torching above the tree canopy. The other part of me felt relieved that the fire was so far away—and could remain, in mind, an abstract entity.

Firefighters from our district were already waiting at the provincial border to slow the spread of the fire as several small towns in the area were at risk of being overcome. Days ago, Forestry had established a hundred-man camp for the wildfire they called ABC001. Every reported and confirmed wildfire gets a name, I was learning. *ABC* stood for Alberta–British Columbia. The numerals 001 denoted that it was the first cross-province fire of the 2016 season. The naming of wildfires was much more than geographical; it was political and, most importantly, economic. Who would be responsible for managing the wildfire? And who would foot the bill?

"XMA685, are you by?" the radio suddenly barked.

I jumped at the sound of my neighbour's call sign, XMA685. His fire tower was roughly fifty kilometres to the east; we both looked out upon the Métis farming community. But who was calling? It was Channel 99, the radio frequency reserved for lookouts to discuss potential smokes and get cross-bearings from their neighbours in order to gain a precise location before reporting it to dispatch. Pilots and firefighters couldn't tune in to this frequency.

"I'm seeing smoke over the farmland to your northwest," the mystery voice boomed over the radio.

The word *smoke* nearly sent my coffee flying. That's impossible, I thought. Moments before, I had been looking north and hadn't seen the faintest sign of smoke. I spun around and was horrified to see that there wasn't just one smoke, there were *four fucking smokes*! I grabbed my binoculars off the Fire Finder and focused on the largest column. There was no mistaking the smoke column for anything else. "You'll know smoke when you see it," a seasoned lookout had told me before I flew out a week ago. "You'll know it the way you know the difference between food that's good and food that's gone bad."

The fires must be forty or so kilometres away. What could it be? Fires burning in a farmer's field? Power line fires? I couldn't think straight. The smoke gathered and rolled white, much whiter than I thought smoke could be. My hands shook as I watched it billow and expand. The largest smoke column doubled in size, rising high into the atmosphere. Later, I'd learn that the four fires were caused by sparks cast by the wheels of a train screeching along the tracks, which ignited the bone-dry dead grass. It turns out grass fires burn white, not black, and can move up to twenty-two kilometres an hour.

I knew what I needed to do—take bearings on the smokes and jump on the radio to alert dispatch—but I remained frozen in place, binoculars glued to my eye sockets. I could see a handful of farmhouses less than a couple of kilometres away from the smoke plumes. I began to panic. Was this my fault? I should've seen the smokes sooner. In just the minute or so I'd been captivated by the wall of smoke approaching from the south, a new fire was born to the north, and it was now escalating rapidly.

The largest column was massive now, rolling like a wild animal, rising up to a staggering height.

"26, this is XMA685 with a pre-smoke," said a male voice over the radio. It was my neighbour, the lookout to my east, who'd worked at fire towers for more than twenty years. His voice sounded anxious as he reported his bearing and distance on the largest smoke column. "There are four smokes," he said in a panicked voice. "And they're coming up fast!"

The radio lit up with a cacophony of voices. The dispatcher alerted helicopters and firefighting crews stationed nearby, who went racing towards the smokes, now four thick white pillars rising into the sky. I needed to jump on the radio and pass my bearing to dispatch, but I remained paralyzed with fear.

"Confirmed wildfires," said a crew leader from an approaching helicopter. They hadn't even flown over the smokes yet. "We're going to need tankers."

Through my binoculars, I watched the air tankers drop three thousand gallons of clay retardant on the blaze. The blood-red substrate splattered against a sky of white smoke. My sense of incompetency grew as tall as the smoke columns. The fire devoured the dry grass in four long gulps. One of the smokes edged dangerously close to a farmhouse as I watched from afar, shocked. But I did nothing but stand there, watching, feeling terrified—and helpless.

It took them several hours to stop the spread of the blaze, to call the status BH, or being held. It occurred to me that no one had called to ask, "Trina, can you see the fire?" Perhaps they thought, she's just a rookie. It's only her second week on the job. She doesn't have the slightest clue what she's doing.

At 8 p.m., I pulled on my harness and descended the ladder. My breath was ragged. My self-loathing grew stronger at every rung. One hundred rungs down.

Useless.

Useless.

Useless.

Useless.

Useless.

I remembered something one of the rangers had asked me.

"The hardest part of the job has nothing to do with calling in smokes," he said slowly. "What will you do when you're alone with your mistakes?"

"ARGHHHHHHHHHH!" I shrieked into the wind. My own voice sounded unrecognizable. I was physically and mentally exhausted from another long day in the cupola, hot, cramped, and bored, delirious with hunger. My throat was a desert. Frustration spawned, infecting me on a cellular level. For a moment I could see myself, dangling off the ladder like a leaf on a branch. Who was this woman hanging off a hundred-foot tower, screaming as though she was coming undone? Why had they hired me? And why—

"*Aroooooooooooo!*"

I quickly glanced below, half expecting to see a wolf snarling up at me.

"Aroooooooooo!"

Holly tilted her snout skyward and howled up her encouragement. Hurry down, Human! she seemed to say. Her torso wiggled back and forth like a worm on the end of a hook. She looked so ridiculously happy, I couldn't help but crack a smile. The dog had waited all day for my descent, eleven hours of sleeping curled beneath the tower and dreaming in the shade cast by the stand of pine trees. Her loyalty awed me to the bone.

My feet touched ground and Holly danced around me gleefully. I rubbed her head, reached for one of her soft velveteen ears. She led me towards the cabin, moving with the energy of a puppy, yowling happily.

This, I thought. This is why people live with dogs. Not for protection from wolves, but protection from the viciousness of one's own thoughts. A soft buffer against the anxiety of failure, disappointment, and being completely alone.

That evening, I patted the empty space beside me on the bed, she happily jumped up, and we slept together, soundly, offering up our regrets to the nothingness of dreams.

The following day, I watched the largest of the four fires, which had grown to thirty hectares in size, be reduced from a mammoth plume of white to a hazy smear on the horizon. It didn't take the firefighting crews long to call UC, or under control, on the wildfire they'd labelled PWF034, although they'd continue to work it—unit crews spread out through the burnt remains, searching through the blackened grass and forest to douse any remaining hot spots—for several more days.

Wildfires torched the province. In Fort McMurray, the Horse River Fire continued to rage northeast, driven by the prevailing southwestern winds towards the Saskatchewan border. And still—no rain. The extraordinarily hot, dry conditions lit fear amongst managers and policymakers. Forestry officials issued a province-wide ban on any kind of fire activity, from starting small backyard campfires to burning massive brush piles on cleared farmland. The use of ATVs—suspected to be the trigger for the Horse River Fire—was banned in recreational areas. Wildfire scientists predicted that the colossal blaze would double in size by the time it reached the Saskatchewan border. There would be no extinguishing it. Only Mother Nature could quell the hungry spread of the flames.

Until the skies opened, we'd be stuck up in the sky on extreme hazard. My stamina for these long, tiring days was fading fast. My confidence, shaky to begin with, was completely shot.

I didn't want to pester my lookout neighbours and become that gun-shy rookie, a fly buzzing in their ears asking stupid questions. Ralph was calling me nearly every day to check in: "How are you keeping over there?" I didn't feel I was keeping it together. I was falling apart from stress and exhaustion, riddled with anxiety that I'd miss calling in another smoke.

"I'm managing okay," I said. "But this job—it's more complicated than I expected."

"It's not a picnic out here," he chuckled.

I called up the woman who called herself the Happy Lookout. The sound of her voice was so kind, friendly—comforting to my strained nerves. I heard myself opening up to her about my fear.

"Don't worry," she said. "You're doing fine. The first year is definitely the hardest."

The Happy Lookout had worked at the same fire tower for over fifteen years. She knew the preferred routes of the black bears in the area, and the place where the forest orchids bloomed. She planted wildflowers around the cabin and grew salad greens in planters on her screened-in porch. She brought out her own water filter system to harvest the rain for drinking water, and tried to be self-sufficient and live as softly as possible on the clearing of bush where she returned every year.

The woman told me that, in her rookie season, she'd accidentally called in a "spook," the name given to a column of moisture rising from the earth after a hard, heavy rain that looks eerily, "spookily," like smoke.

"Then I called in another smoke. Then another. I didn't realize they were spooks."

A pilot flying in the area had responded on the radio, for everyone to hear, arrogantly admonishing her. "Didn't they train you to tell the difference between smokes and spooks?" he'd asked.

She was completely humiliated. "I couldn't stop thinking about it for days!" she said, laughing woefully.

"Here's the difference I've noticed between men and women," she said. "Men will automatically call in a smoke. But women, we're taught to second-guess ourselves. We wait. We watch. Is it a smoke or not? We worry more about calling the shots. Women tend to doubt their own judgment."

The key to the job, she told me gently, was to learn to trust myself. "Trust your gut," said the Happy Lookout. "And don't worry about making a mistake. That's what we're out here to do—look for smokes and call 'em in fast. Better safe than sorry."

CHAPTER SEVEN

In late May, the rains finally came after two weeks straight of extreme fire hazard. Time stopped completely. I stood on the front porch, mug of steaming coffee in hand, and watched, dumbstruck, as rain poured from the skies. The earth breathed a sigh of relief. The birds sang their gratitude to the heavy clouds. I glanced over to my seeded garden beds and sensed the worms stirring in the soil.

"Rain," I said to nobody. "Rain hard."

The droplets made a song on the tin roof. I closed my eyes. I was back in Uganda, standing under our front verandah with Akello, fingers intertwined. After we planted the garden, dropping corn and beans into the dimpled earth, I remembered wishing for rain. I wanted the rain so fiercely, Akello joked that I willed it from the skies with my eyes.

The rain pattered down into the hollow drums, forming shallow lakes at the bottom of the barrels. It was an enormous relief. A shower at last. I couldn't wait to wash my hair with hot water and shampoo that smelled like wild mint. It had been over two weeks and my hair was heavy with natural oils, so greasy that my blond locks had turned a mousy brown. I was so happy, I wanted to cry.

The sound of rain on the roof meant: Stay inside. No climbing today. Nothing will burn. Boil water for a shower. Dream deep.

I crawled beneath the warm quilt that my mother had stitched for me, Holly curled beside me, and let my sun-strained eyes fall into a much-needed sleep. Rain had never sounded so lovely.

A week later, the trees wore a day-old, newborn green colour that absorbed the morning light so the leaves appeared more yellow than green. The wind kicked up and the leaves rustled; that was how they earned their name, trembling aspen. Aspen leaves shake to moderate the amount of sunlight they absorb. It also helps them capture carbon dioxide from the atmosphere.

"Delta Lima Oscar," I said in my a.m. weather report to describe the newly birthed greenery. Code for Deciduous Leaf Out, one of the official "green up" stages that lookouts are responsible for observing and reporting. Over the early weeks of the fire season I'd report on a succession of green up stages: snow gone, green grass transition, deciduous open bud, deciduous leaf out, green grass stage, coniferous open bud, and coniferous needle flush. For the lookouts who stayed late into September and October, witnessing the full death of summer, the colour would fade from the forest's palette. Cured grass stage was the final curtain call. Then snow.

But I'd discovered that I was observing far beyond these scientific stages. I saw everything near and slight, such as the way the rain brought life to the leaves of wild blueberries carpeting the sloping yard. They weren't green, but rather the same colour as henna ink, the red that women in India use to paint their hands. Everything surprised me, genuinely. Dandelions proliferated. Pollen-drunk bumblebees buzzed from flower to flower, moving with the aerial grace of blimps. I noticed a strange-looking pollinator extend a long, curled proboscis into the frothy head of a dandelion. It seemed to belong in the hidden depths of the Brazilian rainforest, not the boreal. A feathery-winged, aptly named hummingbird moth—*now you see it, now you don't.*

When you have nothing to do, no one to talk to, and nowhere to go, what else is left?

You sip your coffee. You watch for the minutiae of what's wild, those processes beyond your control. Keen observation became a balm against my loneliness. Noticing anything new, however faint, somehow made me feel less alone.

By mid-May, I hadn't yet seen a single lightning strike. The lookout to my east had recently informed me that the first lightning never came later than his birthday on May 18, and he'd been working as a lookout in the North for nearly twenty years. He seemed to be the kind of person who took meticulous records—noting absolutely everything, including lightning strikes and sightings of groundhogs, bears, and wildflowers.

The clouds fattened themselves up on a diet of heat and humidity. The harmless tufts of cumulus, parading by like a pod of seals, had disappeared. After the rains we'd swung from low to high, and then right back up to extreme hazard. On extreme, the familiar feeling of dread returned, and my senses were glued to the horizon.

I heard them before I saw them—a sound no louder than the hum of my solar fridge. I stopped breathing and listened hard. The thrum of the engine, barely audible, grew louder. A mosquito echoing in my eardrum. I picked up my binoculars and searched for the helicopter on the horizon. *There.* The machine, tiny as a black fly, a speck, hovered over a cluster of aspen.

Last night, the duty officer had called to let me know that a firefighting crew was coming to "man up" at my tower for the day. On high and extreme fire hazard days, particularly on windy days when fires can take off rapidly, the duty officer dispatched firefighting crews to strategic locations in the forest, at towers, airstrips, and fuel caches. They spread out on the landscape to be able to respond quickly to any flare-ups.

As the helicopter approached, my heart beat furiously. I hadn't seen people in nearly two weeks, and the prospect of facing them left me with a strange combination of excitement and dread. On the one hand, I was elated to see another human being. On the other hand, I was starting to grow comfortable in my isolated geography. I had fallen into a routine. The space was becoming mine and the fear of what lurked beyond the edges of the bush was fading away. But now the fear of those who occupied a place in civilization, the real world—*out there*—was something new to negotiate. I wondered what the crew were thinking as they drew closer to the fire tower. They probably thought I was a weirdo. A hermit. A social recluse. A spinster. Or, even worse, a Rapunzel.

Then again, they probably didn't give a shit about me, or anyone, or anything other than being dispatched to a wildfire. With a high probability of lightning today, they were likely pumped up, hopeful, and eager to chase smokes. They'd come because they were following orders.

A bright-yellow helicopter burst into my world, cancelling out everything else—the birdsong, the collective workers' hum of the bumblebees below, my own thoughts. It circled my tower in a wide loop. I waved to a crew leader in the front seat, but I couldn't see his expression.

I glanced below to see Holly perked up, ears alert, sitting on her haunches. Her black-and-white-tipped tail wagged back and forth like a windshield wiper.

The helicopter landed at the airstrip. Holly listened for the slowing of its rotor blades then tore away from her spot beneath the tower, her husky blood charged as she raced towards the airstrip. The crew, dressed in their bright-yellow uniforms, tumbled out of the helicopter. I picked up my binoculars, positioned myself behind one of the cupola walls so I wouldn't be seen, and watched them curiously.

I assumed that lookouts were regarded by the wildfire community—firefighters, rangers, pilots, radio dispatchers—with a cautious DON'T

FEED THE BEARS attitude, as though we were feral, emotionally volatile creatures who could easily attack if provoked. We were different, that much was obvious. Working for the system, yes, but working remotely, separately. Loners.

I didn't want to be the lonely woman in the fire tower—but here I was, creepily watching the firefighters through 10-by-42-millimetre binoculars, reinforcing every lookout stereotype in the book.

- ✓ Lonely
- ✓ Strange
- ✓ Desperate for company
- ✓ Sexually frustrated
- ✓ Eccentric

Holly had absolutely no reservations about the men who had seemingly fallen out of the sky. She made the rounds, bounding from one yellow guy to the next. She flopped down next to a firefighter lying on his side. He scratched her belly and put an arm around her, and I felt my hard heart split wide open. One of them broke out a deck of cards. Another pulled out a Frisbee from his backpack. A bearded fellow rolled a black-and-yellow drum of jet fuel towards the helicopter. I could hear their muffled voices and their laughter. So close and yet so far.

So badly, I realized, I wanted to be on that airstrip, playing a hand of cards, chasing after a Frisbee, sharing in their camaraderie, or even curling up next to that nameless firefighter lying in the grass. I entertained the thought for a millisecond before swatting it away. I felt a surge of guilt, thinking of Akello, a world away, who was probably deep asleep right now, curled up beneath our mosquito net.

They spent the rest of the afternoon on the airstrip. I had zero interest in reading, or writing, or strumming my ukulele, and instead stared out the window, watching for smokes, and watching the firefighters. I felt as though I had been thrust into a wildlife documentary narrated by David

Attenborough. *Watch the yellow creatures exploring their new-found habitat.* They bounced from one activity to the next. They went into the bush to urinate. One of them slept beneath the shade of the helicopter. The pilot set up a folding camp chair by the helicopter and read a book.

When the sun dropped low over the horizon, they stood and stretched and one by one piled back into the helicopter. They circled my tower in a final act of bravado, the pilot flying even closer.

"Don't go!" I wanted to say over the radio. "Take me with you!"

After their departure, I gazed longingly at the airstrip, conjuring up their ghosts. Had they even come? The silence was deafening.

I climbed down at 8 p.m. as Holly yowled and danced below me. Hurry up, Human! As I unclipped from the fall arrest system, I noticed that somebody had tied a piece of neon-orange flagging tape to her collar.

"What do you have there, girl?" I mused.

I looked closer at the tape and realized there was a folded-up note attached to it. Intrigued, I opened it. "Text me," someone had scribbled in blue ink. There was a phone number.

I stared at the note, incredulous. Then a flash of anger. Was I the butt end of their joke? A bunch of bored, lonely, horny men, manned up in the bush all day, waiting for action, with nothing but wildfire and sex on their minds. They must've wondered, would she take the bait? Whose number was it? The pathetic woman alone in the woods. Fuming, I crumpled up the note and stuffed it in my pocket.

"*Pffffittt,*" I scoffed to Holly, then laughed loudly, a staccato laugh, because the note was something new, a bit of stupid excitement tossed into the long mundanity of my days.

"Points for creativity, I guess, hey, girl?" I said to the dog, who wagged her tail.

But later that evening, I couldn't stop thinking about the note. Who wrote it? I wondered. Why hadn't he left a name? I thought about a few of the men I'd met at the orientation. Had it been that firefighter, Jay? I began to fantasize about the idea, while pushing pork chops around in the frying pan.

But surely, I dreamily mused on, a rugged, good-looking fellow like Jay wouldn't be interested in a woman like me. I became hyper-aware of my greasy, unwashed hair and my crooked nose. I looked down at my hairy legs, mottled with dark bruises from banging the rungs of the ladder as I climbed, and peppered with red, swollen mosquito bites.

The fatty pork chop sizzled in the pan. Hot grease spat at my arms, scalding my skin.

"Shit!" I shrieked.

Holly watched me from her place on the floor, hungrily licking her lips. Her brown eyes queried me desperately. She never hid what she wanted.

I was disgusted by my desire. Part of me wanted to text the number, to find out who was on the other end, to—for the briefest of seconds—distract myself from my worries, my isolation. But then I thought about Akello and everything we'd built together, and I felt deeply ashamed. I tore up the note before memorizing the number.

Storm activity was blowing in on the southwestern winds.

I gawked out the window, watching the harmless tufts of cumulus grow into towering, black-bottomed beasts. They were juggernauts, battleships, their mushroomed tops so white they were blinding. The wind quickened, sending the anemometer spinning like a whirling dervish. The top of the thunderhead started to flatten out, as though a great hand were forcing it down. Virga spilt, suddenly, from the bottom of the cell. Frantically, I paced the cupola.

A bolt of white, molten-hot light stabbed at the earth—*down, up!* Adrenalin charged through my body. My nerves screamed, *ALIVE, ALIVE, ALIVE!*

"Holy shit!" I yelped.

I spun the Fire Finder around in the direction of the strike and recorded the bearing. Lookouts were responsible for reporting "first strikes," the first lightning strike we witnessed from storm clouds. In some ways, it was an archaic practice. Forestry also used satellite light-ning detection technology, so wildfire managers could simply look at computer screens to see where lightning had struck. A single storm could produce hundreds of bolts of lightning. The lightning maps looked like a Jackson Pollock painting, a smattering of red-and-blue dots to indicate where positive and negative strikes had pierced the forest. But the human eye was still more reliable than technology at identifying the category of lightning. I scribbled down the bearing, dis-tance, and type of strike: wet or dry. "Dry strikes"—lightning strikes without precipitation—were particularly important to spot, as wildfire ignition was more likely.

"567, go ahead with your first strike," said the radio dispatcher.

"Bearing of one-nine-five degrees and three-zero minutes," I recited, steadying my excitement. "Approximately one-eight kilometres. Looks wet under light precip, over."

The heavy storm cloud pulled apart and diverged around my tower. Colossal purple curtains of rain and raging energy drew closer, only ten kilometres away. Thunder cracked and boomed. I felt as if I were being surrounded by a herd of elephants, something that had actually happened when Akello and I visited Murchison Falls National Park, a protected wildlife reserve in northern Uganda. A grandfather elephant with an entourage of four younger male elephants crossed the road in front of our vehicle, forcing us to stop. They could have charged our vehicle, smashing the glass with a single swing of their colossal trunks. I remember holding my breath in terror, my body instinctively pulled back. Akello looked over and laughed.

"Where do you plan to go?" he jokingly asked.

As I watched the slow march of the storms, I spied a strike shooting ahead of the cloud.

"Dry strike," I reported over the radio.

A single bolt of lightning can reach 28,000 degrees Celsius, triple the heat of the surface of the sun. Approximately ten people die in Canada every year from lightning strikes. When lightning strikes a tree, extreme heat vaporizes the sap. The hot steam explodes, splitting the bark, and can potentially catalyze a wildfire.

I spun around the cupola, unable to keep track of the spearing light. I lost count after fifteen, twenty, thirty strikes. Lightning bolts struck the earth within kilometres of the tower, sizzling neon pink, blue, and white. A few of the strikes seemed to pulse with light, forking and splintering across the blackened sky. Thunder crashed in my ears and rain lashed the cupola windows.

"Holy FUUUUUUUUUUUUCK!!!" I screamed even louder into the echo chamber.

The radio began to crackle with pre-smoke reports. The lookout to my south called in a smoke, and several minutes later Ralph reported one. Wildfires were igniting, seemingly, everywhere—a phenomenon Forestry called a "multi-start," when multiple wildfires would simultaneously break out in the same day. Firefighters were being dispatched across the forest. It was akin to what Ralph had once described as "a game of Whac-A-Mole!"

I scoured the horizon for any sign of smoke, but couldn't see anything. Steady flashes of white lightning punctuated the purple sky.

After a blue curtain of rain tracked east across the south, I noticed something faint and black rising up out of the trees: a scribble of charcoal. Was it smoke?

No.

Wait.

Yes!

I swung the Fire Finder around, lined up the smoke through the crosshairs, and scribbled down the corresponding bearing. I gauged the distance—maybe twenty-five kilometres—and grabbed the mic.

"Pre-smoke," I said, my voice shaking. "It's black and coming straight up."

I paused, then spied something orange—*flames?*—dancing above the treeline. Oh my god, I thought. The fire was already torching the treetops, crowning above the canopy.

"567 continuing," I stammered over the radio. "I can see orange!"

And then, just as soon as I'd spoken the words, I realized my stupid mistake.

My mind raced back to one of the images of false smokes they'd showed us at training. It was true, I'd seen and reported smoke. But this smoke wasn't from a lightning fire, it was from an active flare stack at a nearby oil and gas site to my south.

Too late now. A helicopter with a firefighting crew had just been dispatched from an existing wildfire to my smoke detection. I dreaded what I knew they'd say over the radio, broadcasting my mistake to the duty officer, other lookouts, and firefighters working in the area.

"Yeah, we're not seeing anything here," said the crew leader. "There's a few active flare stacks, though. That's probably what she's seeing."

Once again, my poor judgment had dispatched a costly resource to an imaginary wildfire. They'd never want to fly to the site of one of my detections again. I crumpled up the pre-smoke report and tossed it on the floor.

The phone rang, probably the duty officer or head dispatcher, calling to reprimand me for my error in judgment. Reluctantly, I answered.

"Hey there, neighbour," said Ralph. His voice was bright against the raging storm. "Did I ever tell you about that time I called in bird shit on my window?"

A friend put me in contact with a Canadian woman who'd supported her Ghanaian husband through the immigration process. Similar to our story, they'd met in Ghana when she travelled there to work with a local development organization. They fell in love, got married, and managed to secure a tourist visa for him to travel back to Canada. Once they hit Canadian soil, they immediately applied for inland spousal sponsorship.

"We didn't mention our relationship for the tourist visa," she told me one afternoon, as we sat outside a café in downtown Edmonton sipping chai lattes. It was the summer of 2015, before my first season at the tower, the year I stumbled through numerous jobs and struggled to re-adapt to Canadian culture. Throngs of commuters walked by. The traffic blurred behind her. "For sure, they wouldn't have let him in the country."

She told me everything they'd had to overcome with immigration: the mountain of paperwork and bureaucracy, the long waiting game, the terrible uncertainty. They were able to live together in Canada as the government processed their claim, but her husband wasn't legally allowed to work.

"It was really, really hard in those first days," she said gravely. "He couldn't find work and became depressed. He hated the cold weather and missed his friends and family. He started spending money like crazy, which was out of character. But I think it was a way of trying to have some control over the stress of the situation, of feeling like he didn't belong in Canada."

I could sense the woman's exhaustion as she spoke. Her words felt heavy to me.

"We fought," she said, nodding her head. "We fought a lot."

It took two years for their claim to be approved by the government. Recently, her husband had found a job at a meat-packing plant, which he enjoyed, and it had helped to ease some of their tensions at home. Her face brightened as she told me this, although I could sense there was more to the story that she wasn't telling me. Finally, as we finished the last sips of our tea, her composure broke.

"Look," she said, "I'm not going to sit here and lie to you about how hard immigration is going to be. If you asked me today, would I do it all over again? Knowing what I know now?"

Her eyes bore into me. I felt unnerved by her gaze and her candour.

"I would say no," she said definitively. "I wouldn't have gone through with it had I known how hard it would actually be." Then she shrugged. "But that's love for you."

I walked home from the café, hands stuffed in my pockets, studying the cracks in the sidewalk, the woman's words ringing in my ears. I'd hoped she'd offer advice for circumventing the immigration system. Instead, she'd given me a strong dose of reality. We could apply, wait five years, and be denied. Or what if Akello came to Canada and he struggled to adapt? Of course it wouldn't be easy to integrate. It wouldn't happen overnight. I tried to imagine what my life would look like five years from now, working to support our family. I'd be in my early to mid-thirties by then. How would I support a child and a husband, who would have to upgrade his schooling to work as a mechanic? How would I be able to pursue my career as a writer? Families make all sorts of circumstances work, every day, I told myself. I loved him, but I couldn't see how I could balance everything—being the breadwinner, a mother, and the writer I wanted to become—without burning out. I wasn't sure if love would be enough.

CHAPTER EIGHT

As the summer wore on, I swung between emotional highs and lows, exhilaration and depression. It was reassuring to know that my daily wage was filling up my empty financial reserves, and I was successfully supporting both Akello and myself. This arrangement was necessary in order for Akello to keep our apartment and cover major expenses, such as Internet data, so that we could stay in touch. Every couple of weeks I transferred money to my parents, who made bank transfers to our joint bank account in Uganda.

On bad days, lonely days, long days in the cupola, my brain frying from excess sun exposure, I grew resentful of our arrangement. It was less about the money than the dependency. How would I support myself in Canada, covering rent, food, and gas, while supporting Akello in Uganda? For the next three to five years? How would I afford plane tickets? What kind of employer would hire me? How could I afford to write books?

On bad days, I resented him for putting me *here*—festering alone in the bush, feeding the conditions for an emotional breakdown, for a full-on inferno.

"You *chose* to be here," I scolded myself. "You *chose* Akello. This is the life *you chose.*"

I'd climb down at the end of these long, excruciating days, crack open my laptop, and try to call him on Skype. Often it would ring—*beep-beep-beep*—sounding off into the invisible abyss, and I'd be left wondering where he was, what he was doing. Maybe there was a power outage. Maybe his laptop battery was dead. I worried about worst-case scenarios. What if he'd been robbed by a gang of street kids? Those things happened frequently. Before I'd left, we heard about a Ugandan man getting his arm chopped off for his wallet and a DSLR camera.

We went days, even weeks, without speaking. It didn't faze Akello; he seemed to accept the circumstances of our love with strength. Though my physical body had never been stronger—from the one-hundred-rung commute up the vertical ladder to the sky and the hard labour of basic survival at the fire tower—my heart felt weaker by the day.

And then, one day, I got through to him. His face appeared on my computer screen, a hazy blur of the man I remembered, but I could see that he was not smiling, nor laughing.

"My mother died last night," he said gravely.

"Oh my god," I cried. "What happened? Are you okay?"

His mother had barely raised him, and he hadn't seen her for nearly ten years. She had struggled with depression for as long as he could remember. I'd never met her, but I'd been hopeful Akello would one day be reconciled with her. I knew he loved her, although I sensed it was a complicated kind of love.

The Skype connection began to deteriorate. I leaned in closer to the computer screen, as if that would help me make sense of his grief. I managed to piece together the broken call, as though he'd sent me a telegraph, as though I were decoding a message from someone lost at sea:

. . . sick . . . long time . . . wasn't taking . . . medication . . . travelling . . . north . . . funeral . . .

Then the call cut off completely. I stared at his handsome face, a sorrowful expression frozen on my computer screen. Tears streamed down my face. I hated what our relationship had become, these broken connections, financial transactions, and an ambiguous future. Not knowing when I'd see him next, let alone when they'd let him come to Canada. I couldn't hold his hand or lie next to him and rub his back. I had no way to help carry the burden of his grief.

I'd never felt so far away.

Akello's mother's death wasn't the first loss we endured separately. Shortly after I'd come back to Canada, my grandfather, whom I'd been close to most of my life, died, and grief overtook me.

My grandfather had been a writer and a woodcarver. We were always naturally aligned in our passions. There was a soft, almost spiritual closeness between us. He wrote me emails whenever I'd travelled in the world, but I stopped hearing from him when I moved to Uganda. I worried that he didn't approve of my engagement to Akello, particularly the colour of his skin. Defensively, I stopped writing too: a silent protest.

And then, days before his ninety-second birthday, he died suddenly.

We drove to Saskatchewan to scatter his ashes on the family's old farmstead, a quarter section plot outside the town of Wolseley. The land no longer belonged to our family, and it had been left fallow and grown unkempt.

At the memorial, my younger cousin Oscar played a melancholy Irish folk song on the fiddle. I tossed a handful of my grandfather's ashes beneath a caragana bush. He had harvested the caragana for carving walking sticks, using a wood burner to etch elaborate designs into the wood. On my twenty-sixth birthday, the year before I travelled to Uganda, he gifted me a walking stick with a bear, moose, gopher, wild rose, and lily burned into the smooth caragana limb. At the top of the walking stick he had drawn a woman with long hair that flowed down the bough and placed a cross-section of deer antler on her head. "For the queen of the forest," he'd said.

Several weeks after he died, I stumbled upon his emails, dozens of them, inquiring about me and my fiancé and our life together in Uganda. They were tender, heartfelt, inquisitive. He'd accidentally sent them to an old email account, one I hadn't used for years.

My heart shattered. I'd allowed communication to fall away based on my own insecurities, and lost the opportunity to connect with my grandfather in the final months of his life. He'd wanted to know Akello—and I knew that he would have helped me understand the doubt that was now brewing in my heart.

Akello and I weren't able to connect and heal together over the loss of his mother and my grandfather, events that transformed us, as the deaths of those we love do. And though I regretted the borders and systems that kept us apart, I also began to realize we were no longer relying on each other as we once had. A continent of space widened between us.

July 1, Canada Day. Just another long day in the tower, roasting in the heat. The sky was blue and empty of clouds. The day before, dispatch had warned us to stay vigilant for smokes. The number of wildfires spiked on long weekends and holidays, when people were out on the land, barbecuing, roasting hot dogs over campfires, riding ATVs, and lighting firecrackers. A fire could start in many ways on Canada Day. And yet the hours crawled by without any excitement. The horizon was clear, the sky unblemished, and my mind wandered. I wondered what my friends were doing out there, beyond the black spruce and the blue swell of distance. They were probably enjoying themselves at barbecues and long-weekend getaways to nearby lakes—camping, fishing, wakeboarding, swimming. Doing all of the normal things that normal people do in the summer. I felt very far away from that world of silver lakes, of drinking beer and grilling meat and gathering around campfires.

"I was normal once," I said aloud to myself, half-heartedly strumming a few chords on my ukulele.

An American kestrel soared around my tower and perched atop the radio antennae for several minutes. The highlight reel of my day, I thought to myself. I was hungry for a hamburger.

The minutes drifted by on the backs of gentle clouds. I lost myself in the pages of a book, then climbed down, limbs stiff. I shook off my harness and followed Holly's lead, breaking into a run towards the sloping yard, away from the cabin and gardens and our regimented world beneath the fire tower. Movement. I thought of the kestrel performing dramatic circles around our tower and wondered, How often do we move about simply for the pleasure of it?

We moved without a plan. I forgot my bear spray in the cabin and didn't bother to bring my camera. We trotted towards the airstrip, that sweet clearing of bush where the wildflowers thrived and the openness allowed a view from the ground that felt tall, expansive. It had become a daily ritual for us, noticing the small changes. I loved walking through the ankle-deep clover, the big-headed white-and-purple flowers, stirred by my step and releasing a scent that reminded me of riding my horse through the hills when I was a child.

When I was thirteen, I'd saved money from my part-time job at a recreation centre in Peace River, enough to buy Pepper, a gorgeous American Saddlebred horse who stood sixteen hands high. Pepper was a dappled silver giant with a jet-black mane and tail. I was barely tall enough to peek over his back and had to use a milk crate to mount him. We lived in town, so I boarded him at my dad's colleague's acreage, and every day, I rode the school bus to the end of the long driveway and ran to Pepper's paddock. He trotted up to the wooden gate at the sound of my voice. I'd rush to brush him, throw on a bridle, and use the fence to jump on his wide back. I couldn't afford the horse *and* the saddle, so I usually rode bareback. It was also warmer that way, especially in the winter months. Pepper and I would take off at a gallop down a trail through the bush to meet my best friend, Jessica, and her horse, Doc Boy, and together we explored the trails in the rolling hills of the Peace River. When I went to Edmonton to study, I knew I needed to sell

Pepper to someone who had the time to love and care for him. It broke my heart, watching the new owner lead him into the horse trailer and drive away. Being at the fire tower, stirring clover, brought back the memory of Pepper and our adventures together—how free I felt when I was with him.

Holly and I took the trail that cut through the muskeg swamp, leading to the clearing, and every step felt familiar beneath my feet. I rounded the corner, Holly only a few steps ahead. She stopped. I looked down at her, then up.

Oh!

That moment when you peer into a view you've seen a thousand times over and something . . . something is different. In a nanosecond your brain registers a change. What doesn't belong?

I noticed a large hump, a round haystack turned golden by the light. The haystack moved. It was alive. My mind sifted through illogical possibilities. Was it a horse? A moose? But no, the faceless creature was a solid lump. And then the way the light caught the hair, illuminating silver tips, made me realize, very suddenly, very fearfully, what it was. The blood drained from my face.

"Grizzly bear," I whispered to Holly.

She knew. She looked up with those amber planet eyes of hers.

What now, Human? she seemed to ask, nervously wagging her bushy tail.

The grizzly's head was down, pressed firmly into the sweet honeyed clover, devouring the wildflowers. I didn't wait for the bear to look up. I grabbed Holly's collar and quietly backed away, making our bodies disappear beyond the corner, beyond the curtain of white spruce trees. When we were out of sight, I let go of Holly's collar. Feeling confident that the grizzly hadn't seen us, I broke the rule that you should never run from a bear, and instinctively broke into a soft-footed jog, stepping as gently as I could. When we reached the slope, I sprinted for home, Holly galloping ahead. I ran with a giant, terrified grin on my face.

I put Holly inside the cabin, grabbed my harness, clipped in, and sprinted up the ladder.

Up in the cupola, breathless, I cracked open the east-facing window.

There. The bear hadn't budged, and still grazed ravenously in the clover. I reached for my binoculars and gasped out loud. It was the largest bear I'd ever laid eyes on, probably four hundred pounds—maybe more. Definitely. A. Grizzly. It was rare to spot a grizzly in the Peace Country, whereas the more common black bears were practically woven into the landscape. Grizzlies are larger and much more apt to defend their food sources than black bears, which added a layer of danger to my encounter. The grizzly stirred, lifted his massive head, and lumbered towards the very spot where Holly and I had stood. His gait must've stretched several metres. Then the bear stopped dead in his tracks.

He tilted his concave nose to the sky. *Sniff, sniff, sniff.*

He smelled us!

Abruptly, he swivelled, with surprising swiftness, on those huge, fat haunches and strode off in the opposite direction. I watched the golden haystack disappear beyond the forest's edge.

How close we'd come, I thought. What if we'd left the cabin only a few minutes later? What if we had turned the corner and come face to face with the bear? What would I have done? The hairs on my arms stood up, electric, contemplating what dangerous standoff might've ensued.

But we were safe, I told myself. The bear didn't want anything to do with us.

The following day, armed with bear spray and the 12-gauge shotgun and my DSLR camera, we tiptoed back to the airstrip. I stopped at the corner. The coast looked clear.

"Hey bear! Hey bear!" I yelled.

Nothing.

We wandered through the clover to where the bear had stood, and I imagined it bursting from the forest and charging us. Grizzlies can run forty-five metres in three seconds flat before knocking their prey to the earth and going in for the kill by clamping their jaws around the scalp.

I caught sight of a set of tracks in the soft mud that the four-hundred-pound omnivore had left behind and knelt down to place a palm over one of the footprints. It made me feel small and fragile. Skin and bone. The way I'd once felt with Pepper—but this creature was wild, dangerous, *other*. Holly sniffed around, following the grizzly's invisible scent lines, eventually winding up at the spot where the bear had vanished into the bush.

I photographed the animal's tracks. Later, I'd send them to my father, who'd confirm what I already knew to be true: yes, we'd stood less than fifty metres away from an adult grizzly bear.

The encounter with the bear lasted mere seconds. But the sight of those grizzled tips in sunlight, the fear and awe quivering in my limbs, would imprint upon me in a powerful way. I felt lucky to have had the sighting; grizzly numbers were much lower in northwestern Alberta than in the southwestern part of the province. The government had imposed a moratorium on hunting grizzlies in 2006 after their populations dropped into threatened status. Bears were shot by humans, or hit by semi-trucks on highways, or killed on the railways, lured by grain that slipped out of train cars. Or, worse, they were illegally hunted by poachers. Before, I'd only seen grizzlies from the safety of our family vehicle while camping in Jasper National Park. I'd never imagined I'd come so close to one in the wild, and I revelled in the rarity of it, in the deep connection to nature I'd been permitted at the tower.

It's funny how locking eyes with the wild can form a memory, a track on your heart, that becomes eternal.

The rains returned in mid-July and extinguished most of the persisting wildfires in Alberta. Not the Horse River Fire, of course. It had grown to nearly six hundred thousand hectares. On July 5, 2016, they finally called UC, under control, on the fire, but it would take another year to extinguish it completely.

It was hard to fathom such an extreme swing from drought to monsoon, but it shouldn't have come as a surprise. Elsewhere in the world, climate change was wreaking havoc. Melting the tarmac highways in India. Scorching staple crops in Somalia. Flooding riverbanks in Bangladesh. Extreme was becoming the new normal, and in the northwestern Canadian boreal forest, suddenly, it was no longer fire season. It had become monsoon season, and once the rains came, they never left. My rain barrels overflowed. I showered nearly every day and had to dry my laundry inside on racks by the propane heater. I couldn't keep up with mowing the thirsty grass, and my garden beds grew tangled with weeds.

The rain fell out of the sky in various ways: slapping the earth in a sudden downpour, accompanied by thunder and flashes of light, pummelling the tin roof one day, caressing the roof the following day. It bounced off the earth, running downslope. It sprinkled lightly and soaked deep into the ground. Some days it did not rain so much as mist, the fog falling low and smothering the treetops, obscuring the black spruce in the swamp and my tower in the sky.

Twice a day, I emptied water from the rain gauge and counted millimetres of rainfall. The tally for July was 208—more than double the monthly average.

The helicopter couldn't see my tower and safely land because of heavy fog, delaying my monthly grocery delivery by several days. I dreamt of fresh reserves: sirloin steak and mushrooms, bacon and eggs, B.C. cherries and peaches, corn on the cob smothered in butter. Thank god I hadn't run out of coffee. When fresh food and water finally came, it felt like Christmas. I served the ranger and pilot coffee and peanut butter cookies, and couldn't keep my mouth shut, chattering about the train fire in May, the sighting of the grizzly bear, and the heavy rains. I could hear myself talking a mile a minute but couldn't seem to rein in the conversation.

"Fire season is as good as over," the ranger said glumly.

He was hopeful for an export to Fort McMurray, where crews were working in large camps to extinguish smouldering sections of the Horse River Fire.

"It's so wet out there that nothing is going to burn."

The pilot nodded. "It's like flying over the Amazon."

The purpose of my job—to spot smoke and report wildfires—had disappeared with the rain.

I should have brought more books and hobbies, I thought. I could've learned another language by now. Italian. *Un altro gelato, per favore!* If I'd brought the sewing machine, I could've sewed a quilt for every member of my family. How many days had I been on low fire hazard? Ten, twelve, fourteen? What day of the week was it? Tuesday, Wednesday, Sunday? I couldn't say, but what did it matter anyway? Every day was the same as the day before.

My worst fear about the job had been the wildfires and missing a smoke. I began to realize, however, that it was the rain that would drive a person mad out here.

On the days when nothing burned, the radios fell silent and there were no visits from firefighting crews. Lookouts were left to their own devices. And what few devices there were at a fire tower, I discovered. No television. No CBC radio. Limited Internet connection. There were telephones, yes, but what could I tell people? "What are you doing?" "Watching a spider scale a blade of grass," I'd tell them. "Watching the mating rituals of tortoiseshell butterflies."

I took to lying in the dewy grass, watching the clouds above, watching insects crawl over me, reading poetry by Pablo Neruda. I memorized his poems and recited them to Holly, who wagged her tail.

When the rain stopped, the birds sang and the earth released a perfume of divine proportions. I strolled to the edges of the forest and threaded my outstretched hands through the wildflowers: yellow arnica, purple asters, fireweed that grew up to my waist, towering purple lupins. I knelt down and learned the botanical and common names of wild berries, tasting their sweet tartness on my tongue: wild strawberries, dewberries, raspberries, cloudberries, bilberries, and blueberries.

I walked to the trailheads and peered deep into the bush. I was afraid of the bears, but more afraid of the woman I'd become if I didn't leave the yard and wander beyond the familiar.

Holly was game for anything. Always, she led the way.

I slung the 12-gauge shotgun around my shoulder, just in case the grizzly made another appearance, and waded through hip-high grass, soaking my jeans through to the bone, and foraged for flowers and strange mushrooms that resembled flapjacks. Every day, I walked farther along the trail, and my fear began to evaporate. I identified tracks: moose, black bear, and grizzly—probably the same one I saw on Canada Day and hadn't seen since. I nudged a pile of wolf scat with my boot. Holly flushed spruce grouse and squirrels up into the trees and stalked them from the bottom. Their angry chatters echoed through the forest. It was an old coniferous forest dying to regenerate, its floor carpeted in spongy moss and fallen cones, their seeds tightly shut within. Only extreme heat from wildfires would release their seeds to be born anew. Who knew when? Not this summer. Mud clung to my boots as we walked, and lakes formed in the valleys. The air was so thick with mosquitoes that they drew blood by the gallons.

I guzzled water and wandered on and on. Around every corner, at the top of every slope, something new. I found myself hoping I'd meet a bear on the path. I wanted to lock eyes with something wild, something with a beating heart. I'd come to the fire tower carrying my fear—my PTSD—but after months alone in the woods I realized how heavy it had become. I didn't need to be so afraid of the uncertainty, I was starting to realize. Danger wasn't inherently lurking beyond every bend. Wandering through the bush, acutely aware of every scent and sound and the sights of life, made me feel alive and connected.

Holly always led us back to the yellow cabin.

H-O-M-E.

I lay on my back and traced the invisible letters in the sky, and watched a male and female kestrel teach their young how to fly.

I whispered the words of Neruda that had become my own.

What determines the silence
is what doesn't happen,
and I don't want to keep on talking,
for I stayed there waiting,
in the place, on that day.
I have no idea what happened to me,
but now I am not the same.

For so long now, I'd avoided remembering what happened, the secret I'd kept from Akello. But remembering was inevitable on rainy days. I remembered everything until I forgot myself.

CHAPTER NINE

"**Y**ou should never make an important life decision while out at the tower."

I can't remember who said those words. A ranger? An older, seasoned lookout?

In any case, I didn't listen. I broke up with Akello in an email.

After a year of concealing my infidelity and growing doubts from him, everything finally came out. I wrote plainly, and though I was devastated—knowing how these words would hurt him—I felt released from the shame of hiding my feelings from him for so long. I loved him, I wrote, but I no longer had the strength to shoulder so much responsibility and uncertainty in making a life together.

I can't do this anymore, I wrote.

I knew it was cruel to end our relationship in this way—an email. Another layer of shame to take on for being the woman who cheated and failed and broke up with her fiancé with a handful of words. But our communication had already been reduced to broken Skype calls and emails, and I didn't want him to go unknowing for a day longer.

I was afraid I'd lose him, that he'd cut me out of his life, that I'd never hear his laughter again.

But I could no longer imagine a future together—not as we'd planned.

I'd spoken my truth, emptied myself out on the page. I'm so sorry, I wrote, over and over again, knowing that words would never be enough to undo my actions.

Akello wrote back the following day. An email written entirely in all-caps.

"WHAT DO YOU MEAN YOU CAN'T DO THIS ANYMORE?
HOW CAN YOU JUST THROW EVERYTHING AWAY?
WHAT AM I SUPPOSED TO TELL MY FAMILY AND OUR COMMUNITY?"

More emails came, one after the other.

"Small one, come home. I'm sorry about my harsh words, but you've hurt me badly. I will never love another woman like I love you. Just come home, please."

"All can be forgiven. We can move forward together.
I will wait the rest of my life for you."

We talked over Skype and cried together. He tried to persuade me to return to Kabale, but I held my ground. I can't do it anymore, I told Akello.

I wasn't a martyr, I'd realized. I'd grown up my whole life believing in the story that a woman could and *should* do it all. But I couldn't do everything and be everything for Akello: wife, mother, writer, bread-winner, activist, community leader. I'd tried and failed.

I was alone, ashamed, and afraid of a future I could no longer predict.

But I never once wavered from my decision to end the relationship. I finally trusted myself. I'd made the right decision. And despite the shame I felt, reverberating in every cell of my body, I also sensed new space, an opening that signalled relief—and maybe even freedom.

The following day, as I climbed the fire tower, I had no energy and my body felt as if it was made of concrete. I looked down, lungs heaving, and contemplated an ending.

It would be so easy to jump.

Why do they always say that? I wondered. *Jumping*. No one jumps.

There's only letting go.

There's only falling.

Akello had told my parents everything. I received a barrage of angry, shocked text messages from my mother. My parents wanted to talk to me, she said, later that evening, to try to convince me I'd made the wrong decision. Over the phone, they had talked about their own marriage, how they'd overcome so many challenges in their thirty-some years together.

They were proud of me, their humanitarian daughter, and I had disappointed them. They had said the things I expected, and things that surprised me too.

"Well, I guess you don't love him enough," snapped my mother.

"I guess not," I snapped back angrily, knowing she was wrong, and hung up.

My shame grew wings, it became wider than the sky. It set the whole forest on fire.

I looked down and saw Holly asleep beneath the pines, curled up into the shape of a moon snail. Who would feed her tonight if I let go? I wondered. We needed one another. For love. For survival. What was it the pagans said about the origins of her name?

When you are afraid, call upon the Holly.

I hung on—a finger grip on a cliffside.

In the cupola, I called my lookout friend, Sam, who was looking out upon a forest of black spruce five hundred kilometres to the north.

I sobbed uncontrollably and sank down to my knees. At first he said nothing. I'm not certain how long we stayed on the phone, but it was long enough for the fading light to change the colour of the trees to a melancholy blue. When my sorrow softened, Sam finally spoke.

"I'm proud of you," he said.

Ralph had witnessed so many changes at the fire tower over the years.

In the summer of 1959, he recalled sleeping in a canvas tent and climbing, not a tower, but a 10-foot wooden platform, and relying on a makeshift wooden protractor to take bearings on smokes. That spring, he'd butchered a pig and salted the meat, which he hung in flour sacks beneath the platform. Ralph had to chase away the bears, and other predators from his summer-long supply of meat. He was only twenty years old. Before the days of propane heaters, lookouts felled trees, split wood, and fed potbelly wood stoves to keep the cabins warm. They scoured the woods for springs, or streams, where they could procure their own drinking water. Back in the day, bears busted through the front doors and climbed in through the windows to ransack the cabin for food. There were stories of lookouts making narrow escapes and seeking refuge in the tower or outhouse. Today, the windows were built at least a foot higher off the ground to prevent bears from breaking in. When Ralph started on towers, the only mode of communication was the two-way radio. Lookouts used Channel 99 to chat at night, telling jokes, sharing recipes, and strumming guitars—broadcasting their songs into the forest.

But the hardest part of being a lookout, said Ralph, had nothing to do with bears, food, water, nor technology, and everything to do with being human. The hardest part in 1966 was the same in 2016.

"The hardest part of this job is having no control of what's happening beyond the fire tower."

A few days after Akello and I broke up, Ralph called back. He wanted to sing me a song.

He'd written alternative lyrics to a Joni Mitchell song, and had sung it for other lookouts in the province, ones he'd trained or shared the forest with over the years. Several years ago he sang it in front of lookouts and rangers and fire managers at a huge conference in Hinton. It had become an anthem in Alberta's lookout community. "Would you like to hear it?" he asked me a bit wistfully. "Of course," I said.

Without another word, Ralph cleared his throat and launched into song.

> *The clouds they covered up the sun*
> *The thunder wakes up everyone*
> *So many things we might have done*
> *But clouds will rule the day*
> *I've looked for clouds from both sides now . . .*

His voice full of gravel, but tender, reminiscent of Willie Nelson. Warm tears slid down my face.

> *Now my friends are acting strange*
> *They listen to me, they say I've changed*
> *But they should sit up here all day long*
> *And look for clouds this way*
> *The clouds have covered up the sun*
> *There is no rest for anyone*
> *The rain gauge filled with dust again*
> *There are no clouds some days*
> *I've looked at clouds*
> *From both sides now*

Nothing would burn for the rest of the season.

I never stopped climbing up to the sky. It could have felt like a prison cell, a punishment for my sins, but it felt more like a cathedral, a refuge. I wanted to be alone up there, and delay going back to civilization and facing my new reality. I'd have to tell everyone what had happened with Akello, why we'd broken up, why I wouldn't be travelling back to Uganda. I wasn't even sure where I'd go, or what I'd do.

Looking out into the forest was a kind of medicine. Some days I felt a calm acceptance wash over me. It wasn't elation, or devastation. It was a neutral way of seeing nature, a space to forget about projections of good and bad.

To simply be.

And then one day the news came suddenly, swiftly as a tree swallow flying into the cupola. An email from my literary agent in Toronto with the subject heading:

HURRAY!!!!!!!!!!!!!!!!!!!!!!!!

I clicked open the email.

"Oh my god," I said to nobody.

A small but respected Canadian publishing house was going to make an offer on *Women Who Dig*. My book was going to be published. I was going to be an author.

I was going to be a writer.

"Oh my god," I whispered to the trees.

My mind flew back to that day when I was fourteen years old and my dad brought home a copy of Pearl Luke's book *Burning Ground*, the novel she'd penned during the seven years she worked as a fire tower lookout.

"Maybe you could do that one day," he'd said.

"Maybe," I'd shrugged, because it was far-fetched to contemplate becoming a woman alone in the woods, surviving a summer of solitude, let alone becoming a writer—a published author.

My body felt light, as though my bones were hollow. I was so happy, I wanted to fly out over the forest and into the sublime space where the horizon and forest bled together.

Instead, I climbed down the tower that evening, threw off my harness, and danced with Holly.

"ArooooooooOOOOoooooooo!" she yowled.

"ArrrrooooooOOOOOOOOooooooo!" I yowled back.

Our wolf song reverberated throughout the forest.

In the last weeks of the season, I found myself climbing the tower late at night. I watched the sun sink into the earth at nine o'clock and set the sky on fire. I felt my lungs wanting to fly right out of my chest. It was a glorious, holy sight.

On the evening of the full moon, I dragged the old aluminum bathtub over to my firepit and boiled pot after pot after pot of rainwater on the gas stove. I slowly filled up the tub, keeping the heat in with an old blue tarp. After the last pot was poured, the water steamed and the full moon appeared, a wide white face over the eastern horizon, birthed from the teeth of the black spruce.

The death of summer was in the chilled air. Autumn was already here. I slipped into the steaming liquid. Holly guarded the campfire, peering out into the inky night, sensing the presence of wild things beyond the black spruce, creatures whom I no longer feared.

The moon wandered high into the sky.

I stacked my empty bins and packed bags by the front gate, swept out the cabin, put up the wooden shutters, and emptied the last of the rain barrels. I plucked a thin carrot from the earth and said goodbye to the garden beds. The rest I'd leave for the birds.

After 130 days alone, I was finally going back to civilization. I felt a strange mix of nostalgia and excitement as I prepared to leave the tower.

Over the course of the summer, I had fallen in love with a dog. I had been dazzled by acts of nature, witnessing first strikes and wildfires and weathering the sun, rain, wind, and fog. I had nearly come face to face with a grizzly bear. I learned to love spiders and their daring silken adventures up my one-hundred-foot fire tower. I hadn't called in a single wildfire—though I'd witnessed the suddenness of ignition, and the speed at which fires can travel, from afar. I sensed their transformative power on the landscape.

And I'd lost my love, Akello, a big part of my reason for taking the lookout job in the first place.

Yesterday, I had nearly got lost in the woods after diverting from my compass bearing and moving east instead of north. For fifteen minutes, I panicked. Bushwhacked. I worried that a crew would have to come rescue me. Luckily, we emerged from the bush onto an edge of the airstrip. I was so grateful to see my yellow cabin that I fell to the ground and kissed the earth.

"At the end of every season alone," the woman who called herself the Happy Lookout told me late in the season, "we get to be reborn again into the world."

What kind of woman would I become?

Nothing had ever felt more uncertain.

In the distance, I heard the approaching helicopter. The sound was faint, no louder than a hum.

PART THREE

WINTER BURN

The deep ache of this audacious Arctic air is also the ache in our lives made visible. It strips what is ornamental in us. Part of the ache is also a softness growing. Our connection with our neighbours, whether strong or tenuous, becomes too urgent to disregard. Twenty or thirty below makes the breath we exchange visible: all of mine for all of yours. It is the tacit way we express the intimacy no one talks about.

—GRETEL EHRLICH, *The Solace of Open Spaces*

CHAPTER TEN

After the helicopter landed, a wildfire ranger drove me and Holly from the airport to Peace River. The feeling of driving along the highway was exhilarating to the point of terrifying. "Slow down," I wanted to caution him, though he was driving at the speed limit. I hadn't moved faster than the pace of my own making for over four months. Some of the lookouts catapulted themselves straight from the fire tower into their off-season lives in downtown Toronto, Montreal, and Vancouver. I couldn't fathom it.

The population of Peace River was only seven thousand, but I felt as if I'd migrated from an isolated existence into the throes of a bustling metropolis. I already missed the savage calm of my monkish life in the bush.

The ranger took me out for lunch at the town's café and I tried not to stare too closely at the other patrons. So many people doing all of the things that people do: sharing a laugh with a friend, wiping mayonnaise off their lips, scrolling on their phones, cutting up food into bite-sized pieces for a baby, losing themselves in the pages of a book. I studied their facial expressions, which conveyed a thousand nuanced emotions. It occurred to me that I missed gazing into another person's face more than anything else. That is what we take for granted when in the company of people—being able to read one another's faces, studying the curvature of lines, the movement of lips, the brightness of eyes, and placing ourselves on the map of another human being's emotions.

"How are you?" I asked the man behind the counter.

"Good," he said in a single exhale. An enormous smile stretched across his lips. "Thanks for asking, actually. No one ever takes the time to ask me." He shook his head and laughed.

Part of me wanting to reach across the counter and touch his face to confirm that he was real, but I resisted the impulse. I hadn't forgotten about social order and followed the rules like everyone else. I ordered my food. Paid. Sat down at the table, across from the ranger, and carefully ate my Montreal classic sandwich, wiping away the mustard at the edge of my lips.

I could have followed my familiar inclination to leave, but I stayed north in the Peace Country.

It wasn't so much a decision as an instinct, not unlike the way a bear claws beneath the earth for the winter, carving out a home for herself under the roots of an old aspen tree. I wanted to sleep away the season, curl up with my own shadow, feed on grief, and dream of greener seasons.

The truth was that I was afraid to go anywhere else. I couldn't go back to Uganda, though I desperately missed Akello, Patricia, and the beautiful community I'd cultivated there over the past three years.

And I couldn't face my activist friends, colleagues, and community in Edmonton. I was afraid they would no longer accept me.

My parents lived in Peace River, but I rarely saw them. They were still disappointed with me. My mother and I snapped at one another like wolves, and my dad couldn't look me in the eye. I felt the absence of their emotional support, a layer of abandonment on top of the shame. My brother, his partner, and their three-year-old son lived in Edmonton. We'd drifted apart over the years of my travelling and work abroad and missed out on knowing the details of one another's lives through most of our twenties. We hadn't really reconnected in a meaningful way since my return to Canada in 2015, although, deep down, I longed for the protective older brother I'd had growing up. I wished he could tell me that everything was going to be okay.

Akello and I spoke every now and then. The tone of his emails and voice during Skype conversations gravitated between anger and grief.

"What am I supposed to say to my family, Trina?" he asked.

"You have to tell them the truth," I said. "That I'm not coming back."

"That is not how we do things here," he said. "You don't understand our culture."

His words were intended to hurt me—and they did. I winced, reminded of always being an outsider in Kabale. But I also respected his decision to stay quiet about our split. I could only imagine how the naysayers, a few of our Ugandan colleagues who doubted the relationship, would respond to the news. I hated that he'd have to shoulder their cruelty alone.

"With time, people will come to know the truth," said Akello.

I rented a mobile trailer ten kilometres outside town, which perched dangerously close to an eroded cliff overlooking the Heart River, a small waterway that emptied into the Peace River. The landlord was an old trapper and I didn't sign a lease. He said that a handshake was enough.

"I don't yet have a job," I admitted when he showed me around the place. But I was hardly an anomaly in an oil and gas and logging resource town where employment came and went in boom-and-bust cycles. Under the current recession, more than 10 percent of the local population was unemployed.

"I'm more concerned that you don't have a truck," he said gruffly, pointing to the long gravel driveway that threaded through the bush. "Winter is coming soon, and . . ." His voice trailed off.

He dropped a single key into my palm. His skin felt like sandpaper, thickened from prolonged exposure to winter's teeth. For many years he'd set snares along his trapline, unhooking the limp bodies of rabbits, martens, and lynx from where they'd met a swift death. He sold their soft, beautiful pelts for a modest price. The landlord cleared his throat and cast an awkward glance down at my five-foot-three frame.

"Are you going to be okay out here alone?"

"Yes," I answered half-heartedly.

The trailer was nearly off-grid. Fortunately, it had electricity and an underground water tank, and a pump provided running water, although I'd have to pay to truck in water once a month. For heating, I'd rely on a wood stove. There was an old diesel furnace for backup, but the diesel was costly, so my main task would be splitting and stacking firewood to constantly feed the wood stove.

The trailer was poorly insulated, the walls yellowed with age. I moved my belongings into the rectangular shoebox and bought cans of paint that reminded me of the vibrant southern places I'd travelled: turquoise, terracotta, mint, mustard, and periwinkle. I hung up my artwork and tried to convince myself that I'd come home. That I could find myself anywhere in the world.

But on the land, I didn't need convincing. The trailer had been plunked down on the edge of the valley where the eye was free to roam towards the terrific horizon, or down into the cascading hills, settling on the meandering Heart River, a black snake that slithered along the valley floor. My nearest neighbour lived a kilometre away.

In late October, I went walking along the valley edge. The last of the tangerine-coloured leaves had fluttered to their death. The land went on forever. I cut through deciduous forest, noting the bear claw marks dug into the silver aspen, scarred black. I spooked a cow moose loose from the bush and she high-kneed it away through the dense brush. From the forest, I was released onto a long clearing of wild grasses, the whole valley spread out below. I walked through waist-high grass and stirred wild sage that had already gone to seed.

I lay on my back and watched the red-tailed hawks soar above.

Here, I thought. Here is where I can heal a broken heart.

When animals are injured, they slink away into the bush to lick their wounds. Surrendering was easy in the North. There were so many places to hide.

I lay on my back, feeling the presence of something powerful beneath me, under me. Let me stay north for a while, I asked of the earth.

Let me rest
freeze
sleep

here

until geese
thread the sky,
pointing north,
crocuses slip
from the frozen
slant, sap flows
and heat shakes
the dead:
rise again.

I wasn't alone.

How could I let her go? When I returned Holly to the farm in Manning, I knew I'd made a mistake. On the drive back to Peace River,

I kept glancing in my rear-view mirror, expecting to see her eager grin, that pink tongue dangling out of the side of her mouth. My eyes blurred with tears.

I called the owner the following week.

"I'm not going back to Uganda," I told her. "I'd like to keep Holly."

"We were hoping you'd say that," she replied.

I brought Holly home to the trailer on the cliff's edge and we took to exploring the valley together. Holly scented out moose sheds, where the herds of wapiti elk lie down to sleep, and a freshly dug bear den in the side of a southern-facing slope.

At night, I'd sleep with one hand on her warm, familiar body, feeling for the steady rise and fall of her chest. When I left her at home alone, I'd return to discover she'd stolen an article of my clothing—a mitten, a toque—and curled around the ghost of me.

She seemed to need me the way I needed her.

Before the snow fell that year, Jay, the firefighter I'd met at orientation in the spring, came back into my life. His northerly camp had been closed for the season, and Forestry had moved the few remaining firefighters into a hotel in town until October, the end of the late fire season. Soon he'd be released into the "off-season," as everyone called it. Free to wander, travel, explore. The winter months were our summer.

We bumped into one another in town one day. Jay was even more handsome than I remembered—his body thicker with muscle, his beard overgrown, unkempt. He made my face split into a wide grin. I could hear myself laughing, a sound I thought had gone extinct.

I liked him.

My whole body liked him.

And he knew it.

But he seemed so young and energetic. Naive, maybe. Jay was only five years younger than me, in his mid-twenties, but I felt as though the last few months had aged me.

I gave him my number, though I wasn't sure why. Flattery. Distraction. Or maybe—hope.

He texted every day. Trying to get through to me.

"Let me help you paint your house."

"I can split wood for you."

Although he seemed kind and willing to get to know me, I made myself unavailable, hidden away in the woods. You can't start a relationship with a broken heart, I reminded myself.

"I just broke up with my fiancé," I said to Jay over the phone one day, my voice trembling like a leaf.

That's all I could manage to share of my story with Akello. I wanted to speak my truth, but I didn't yet know how. I'd survived the wildfire, but I couldn't make sense of the woman left standing in the ashes, her life half-torched.

"I understand," he said, though he did not.

One afternoon, I let my guard down. Jay and I went hiking together in the hills, trekking the lip of the valley through the tall grasses, picking wild sage and rubbing it between our fingers to release the aroma of spice and medicine. We pressed our backs against the grassy slopes and sipped cans of beer.

"Listen," I said. "Do you hear that?"

We stopped breathing and listened hard. Then, *there*. A sharp shrieking sound echoed eerily through the valley. I thought it might be a hawk or bird of prey.

"It's a male elk bugling," he said. "Calling the females in the area."

Jay looked over at me. I could sense his expectation, his boyish hope, as though he wanted to hold my hand or press his lips against mine. I turned away and stood up suddenly, as if I had somewhere to go. We followed after the elk's haunting love song, weaving our own path through the naked brush, stepping over fallen logs, bushwhacking through dense tangles of wild rose and willow.

"There," he said quietly, leaning towards me, so close I could feel his breath on my neck, as he pointed across the river through the maze of naked aspen.

My cheeks flushed. Heat crept up my body.

Then I saw the bull, blended into the forest, frozen, large and regal, antlers rising like branches. I looked at Jay and something wild—astonished—flickered between us.

"Don't stay here for the winter," he told me earnestly. "You should get out there and see the world."

I've seen enough, I wanted to say, but I said nothing.

I asked him where he was going.

"I'll see where the road takes me," he replied.

He'd loaded up his canoe, packed his car with all of his worldly possessions, and was going to drive west to the Pacific coast. He wanted to soak away the tension of the fire season in hidden hot springs. Or maybe fly south to Central America. I envied Jay's energy and optimism for the off-season. Part of me wanted to run away with him. Start over. Pretend my other life had never happened.

"Come south with me," he said playfully.

"No," I said. "I think I'll stay here for a while."

We embraced. He held me for a few long seconds. Then I pulled away, putting space between us.

"I'm sad to say goodbye," he said.

I'm sad too, I wanted to say.

"Maybe I'll see you next fire season," I said, shrugging, stifling my emotions.

I watched him drive away down the long, muddy road until he disappeared.

———

Snow came hard and fast, pelting against the trailer windows, obscuring the valley. The temperature dropped to twenty below zero Celsius and the Arctic winds ripped up over the ridge, pushing banks of snow against the side of the trailer.

"It's good insulation," the landlord had told me, so I shovelled more snow against the north-facing wall, though it was useless.

I could feel winter's reach prying through the cracks in the trailer walls and on the windows, which were etched with intricate ice designs. I shoved blankets beneath the front door to ward off the creeping frost. The old wood stove wasn't as efficient as I'd hoped. Every few hours I had to lay down more pine and spruce logs on the blistering coals or a terrible chill overtook the trailer. I was living in an icebox.

I acquired three axes for the daily task of chopping wood: a boy's axe, more of a hatchet really, for splitting kindling; a regular axe with a wooden handle for quartering logs; and finally, not an axe, but a maul, a heavy-headed beast that resembled a sledgehammer.

The abundance of firewood, stacked neatly from floor to ceiling, began to vanish into the piercing-cold air. Every day, I loaded the wagon high with split wood, but my supply was dwindling and winter had only just sunk its teeth into the North. The feeling of scarcity shook my confidence. But my unease was about more than the firewood; it was also about my own psychological fuel to endure, alone, following a summer isolated in a fire tower. Although I saw my parents every now and then, and I knew that they loved me unconditionally, I couldn't help but feel the emotional distance—a gulf of hurt and disappointment—between us. I spoke with a few of my closest friends from Edmonton over the phone.

"Come back to the city," one of my best friends urged me. "You can stay with us."

But I couldn't imagine resuming my life and work in the non-profit sector in Edmonton.

I didn't know where home was anymore.

————

Winter's harsh beauty lured me onto the land.

One afternoon, I stepped into my cross-country skis and glided down the long driveway. Hoarfrost clung to the aspen boughs like the velvet on a deer's antlers. Holly leapt ahead gleefully, pushing her snout into the snow and flinging her head back so the flakes stuck to her whiskers. We happened upon the carcass of a frozen doe, her body half-eaten by wolves or coyotes. Ravens dispersed, flapping wildly, as we approached. The bones were nearly picked dry.

When I finally stumbled in out of the cold and dropped, rag doll, onto the sofa by the wood stove, I was too exhausted to remove my ski pants. I pulled a blanket over my body and huddled into my own shelter, but I couldn't shake the cold. It crawled all over my body, permeating layers of useless cloth.

I glanced at the half-empty woodbox. Did I have enough to get me through the night? I thought of the frozen deer carcass, the bits of hair, laced with ice, clinging desperately to what used to be a living, pulsing, running thing. If I died out here, tonight, who would know? How long would it take them to find my frozen body?

My throat began to burn and my head turned heavy as a tombstone. Fever, *rising.*

My eyelids dropped like a baited hook through a hole in the ice and I went

down,
        down,
                down

until I was cut loose on a raft, drifting in the waters beneath the ice. The space was hollow as the insides of a whale's belly, an indigo chamber. I saw myself clinging to the raft wearing a white lace dress that fell to my ankles. We were sinking. The temperature of the water climbing, heat crept from my toes up my ankles and shins and pooled at my kneecaps.

My fingers twitched, armpits leaked perspiration.

My head was on fire.

My body screamed awake: eyelids yanked open. I turned on the sofa, aware of my weak limbs, suddenly, drenched in sweat. My throat had become an open wound. I took a sip of water, but it was like swallowing arrowheads.

I lay in my body's sweat, watching the orange glow through the wood stove window.

The light throbbed and dimmed.

Some wildfires in the boreal forest burn year-round, even on the coldest, darkest days of the year. Wildfires burrow deep in the peat and humus, smouldering, slowly chewing up the dead, dry debris, hidden beneath layers of snow crystals. The Horse River Fire, the wildfire that evacuated the city of Fort McMurray, was asleep, deep beneath the lock of winter—but not yet dead.

Even if a woman climbed her fire tower, bare hands freezing on the bone-cold rungs, to look out upon the haphazardly burnt, charred forest the Horse River Fire had galloped through, she might not see the smoke from the winter burn. Inversions, cold air trapped near the ground by an upper layer of warm air, could force the smoke to stay low to the ground until spring.

One day, while I was standing in line at the café in Peace River, an acquaintance turned around and asked me loudly, "How is your husband in Uganda?"

"He's fine," I said meekly, because I could not bear to correct her, explain myself to a near stranger.

Weeks later, another acquaintance asked about him, and I admitted that we'd broken up.

"Well," she said, "it must have been a good adventure."

Her words felt like hurled stones.

A good adventure.

A friend mailed me a copy of Zadie Smith's novel *Swing Time*, which explores themes of race and power. I feared the worst of myself. Had I become Smith's archetype of the white bleeding heart? She writes about a white, wealthy woman—a pop star named Aimee—who goes to west Africa, plucks a black baby out of her mother's arms, and hastily marries a black African man. I cringed when I read about Aimee's *adventures* in Africa. Was I really Aimee? Had I been blind to my white privilege, and the barriers that existed between Akello and me? Had I been too caught up in my own pleasure-seeking ways, relishing in *a good adventure*?

Someone whom I once considered a close friend answered for me. Publicly.

"She cheated on him. Then dumped him. Then abandoned him."

She posted these words in a comment beneath a photograph of Akello on Facebook.

At first, I wondered how I could defend myself. It was painful but true, what she wrote. I'd done exactly to Akello what she accused me of. And yet—it was also an overly simplified narrative that omitted so much about what had happened between us, making Akello into a victim and me into a villain. Akello wasn't a victim: he was a strong, intelligent, capable man.

Maybe I hadn't been strong enough to hold up my end of the relationship. I'd fucked up—that part of the story was true. But I wasn't a bad woman. I'd tried and failed. And in the process I'd given so much of myself and my love to Akello. Beyond my white privilege and the differences between us, I truly believed our love had been a reciprocal one. I knew that Akello would say the same thing, despite everything.

I didn't have the energy to respond to my friend's accusation. Actually, it was my mother who came to my defence and confronted her. Later, the woman deleted the comment.

"Thank you," I said to my mom.

"You're my daughter," she said plainly. "You made mistakes, but you didn't deserve that."

I wanted to start believing a different story.

I signed the book contract with the publishing house for *Women Who Dig*. The book would be published in the spring of 2018—only a year away, and yet I found it hard to think ahead to that point and imagine what I would be doing. But it was something real in the world, something good and hopeful that I could tether myself to. I dreamt of holding a copy of the bound book in my hands. It was a guiding light during the dark winter months.

I began to feel the land acutely. Holly and I spent hours outside, hiking through the forest, exploring the river valley. Boots wading through powdery snow. Holly bounding through the banks.

I noticed the heart-shaped chaga mushrooms growing on the torsos of birch. A lone coyote zipped ahead of us on the trail. I plucked shrivelled wild rose hips off the prickly red bushes and sucked on their tangerine skin. A single hip had the same amount of vitamin C as an orange. First Nations peoples call them "survival berries" because they cling to life fiercely through the extreme cold. I took to studying tracks in the snow—rabbits, moose, deer, mice. The ruffed grouse made angels in the loose snow with their wings. When the sun dropped below the horizon, the coyote chorused in the concert hall of the valley below—yip-yip-yip-aroooooooo!

One night in late January, I woke up to feel Holly stirring at the edge of the bed. I glanced at my clock: 1 a.m. The dog was staring attentively out the window.

I pulled myself up to the frosted glass.

"Oh!"

Under the eye of the January moon, a herd of wapiti elk—maybe thirty, forty of them—were pawing beneath the crust of snow, grazing outside my bedroom window. They were only metres away from where I slept.

Was I dreaming?

Two bulls gently butted racks and the cows rose up on their hind legs and bullied one another for the best grazing spots. During the day, elk dispersed into thinner herds, hiding in the forest, but they came together in the thick of night for socialization and protection. I thought of the elk's haunting love song and that afternoon of getting lost in the forest with Jay.

If only he could hold me now, I thought, feeling the ache of empty space in my bed.

For hours, I watched their blue-black, shadowy bodies, until my eyes couldn't stay open.

In the morning, I ran outside in my pyjamas and snow boots. The elks' hooves had trammelled down the snow and green grass poked up where they'd pawed away to reveal bare earth. But they were completely gone—dispersed—hidden away in the valley below.

Here's what I learned from that winter alone on the Heart River: you cannot hide from people in the North. It would have been easier to hide in an apartment in downtown Edmonton, Vancouver, Montreal, Toronto, or Halifax. You cannot stay anonymous forever in a small northern town.

The man down the road, my closest neighbour, ploughed my kilometre-long driveway free of snow. Ranger Jim, the same one who had flown me out to the fire tower the previous summer, dropped off truckloads of firewood every month. "Don't worry," he chuckled. "We'll keep you warm for the winter." A family friend made me a beautiful wooden cutting board and another friend dropped off an old couch. One woman invited me into her home pottery studio and taught me how to work with clay, how to embrace the slow, old process of throwing and firing vessels. I found a part-time job as a server at the local café and developed friendships with my co-workers and some of the patrons. I found my voice amongst people again. I rediscovered that, deep down, I wanted to belong to the world.

Slowly, my parents and I began to repair our relationship. They drove out to my new home on the edge of the valley and we shared meals together. They fell in love with the land around my home. We wandered along the edge of the cliff, boots crunching through snow.

"Oh!" my mother cried, standing on the edge of the cliff, moved by the beauty of the Heart River valley. "This is incredible!"

I knew she communicated with Akello now and then, but I didn't mind. They had built a significant relationship over the years we were together, and the loss had affected my parents too. We didn't talk about the painful conversation that happened while I was at the fire tower last summer, but I decided to try to let it go. I knew they'd been stunned by the news.

"I can see why you made the decision you made," my dad told me in January. "You guys were facing a lot of challenge and uncertainty. I think you made the right call."

It was a relief to be seen and accepted by my parents again.

My brother called and texted more often from Edmonton. Slowly, we were rebuilding our relationship. I could feel his love, even from afar. He was really trying.

He confessed that it had been hard for him, all those years I spent travelling and working abroad, dealing with my absence.

"I worried you wouldn't ever come home," he said.

I thought of my brother as I'd known him as a child: a hockey star confidently chasing the puck, or showing me how to build a campfire, or swing a hatchet and split wood. I hadn't realized that he'd needed me too.

He and his partner were pregnant with their second child. A girl.

"I want you to be in her life," he told me. "My kids need to know their auntie."

I'd forgotten how much I missed my brother.

In February, when my firewood reserves dwindled, a group of new friends congregated at my trailer to help split and stack firewood. They brought lasagna and bison roast and salads and homemade bread, and we feasted, gathered together on the floor, sprawling around the warmth of the wood stove. We walked to the valley's edge and I shared with my friends the expansive view below. I watched their eyes grow wide with wonder, their jaws drop open. I listened for the collective loss of breath, the sound of exalted surprise, of being moved by the beauty of nature.

Some views are better shared than witnessed alone.

At the Heart River, what surprised me the most was the feeling that came in the dead of winter. On the coldest, heaviest, darkest day. It was stronger than a feeling. A haunting, perhaps.

I couldn't shake the sensation loose.

I missed the fire tower.

I hungered for that faraway home in the black spruce and swamp, and for the weather and elements that dictated the extremity of my days: sun, lightning, wind, and fire.

I couldn't deny the longing to wander deeper into my isolation. Why? I was already alone. I was also toeing the edge of community. Welcoming people back into my life.

Returning to the fire tower, after everything, seemed equivalent to voluntary exile.

Did I want to exile myself again?

Spring crept slowly towards the North.

Every day, I noticed the small changes of the coming season. Snow on the south-facing slopes disappeared. Crocuses blossomed from the half-frozen ground—small purple flowers that rose silently, subtly, from beneath the dead grasses. Stumbling upon a crocus, I dropped to my knees, cleared away the dried grass, and stroked its small, fuzzy buds. When the sun dropped low over the western sky, the soft petals glowed, soaking up the dying light.

My heart quickened at the sound of geese, the faint honk of their exhausted song. They, and millions more, were travelling north to their breeding grounds in the boreal. Some birds would flap as far as the Arctic tundra, pulled towards the magnetic land of freeze and thaw.

In early March, an adolescent kid set fire to a grassy slope at the edge of Peace River. The grass fire crackled alive and surged on the back of the wind, scorching the hillside. The local fire department and Forestry both arrived to put the fire out. The young arsonist fled the scene.

"Fire season is officially here," pronounced one of the wildfire rangers.

The following week, I drove past the extinguished fire and saw green blades of grass poking up through the black, ashy remains. How badly the land wants to burn, I thought. Fire cleanses the earth, not unlike rain. Fire in the boreal was as essential as rain in the Amazon. Excessive heat opened the pine cones, birthing new coniferous forest. Fire-loving insects swarmed into the black. The nitrogen remains of fire—a soft layer of ash—nourished the growth of mushrooms and wild grasses and flowers and berries. Fire created food, tender green shoots, for ungulates and omnivores. The arrival of fire signalled birth, not death. And yet, I wondered. How quickly society rushed to extinguish the flames, to prevent the spread of a natural process, to protect homes, businesses, and factories—built ever more rapidly into wild spaces—while also stifling one of the forest's oldest desires. The act of wildfire suppression was leading to the creation of tinder-dry forests, volatile "fuel loads" practically begging for a spark of light. Everyone swatted away the irony that the more humans hurried to quell the flames, the worse the wildfires would become. And yet there were so many values at risk on the boreal landscape—people and properties to protect. Managing wildfire was a hugely complicated task.

Everywhere I looked, I saw signs of spring, the early stages of green up, that signalled the coming fire season. I swore that I could feel the heat in my bones. Time to pack. Time to fly away to my cabin in the woods.

I drove south to Edmonton to stock up on food rations for the summer and packed my bags. I remembered the long days of monsoon—*cabin fever*—and packed extra books, a sewing machine, a bin of colourful cloth, and patterns for simple patchwork quilts.

Holly, sensing that change was coming, sniffed anxiously at the half-packed boxes and bins. She became my shadow, glued to my every movement. I scratched her velveteen ear.

"Don't worry, girl," I soothed her. "I'd never leave you behind."

The phone rang in late April. My tower would be opening up a week early, said a ranger. I counted the days: less than a week away. *Going in*, I scribbled on my calendar. May 8, 2017.

Before flying in, I gathered with some of my new friends at the local pub in Peace River. I soaked up their company, already anticipating the coming season of isolation.

On a napkin, they scribbled down nonsensical things I should do at the fire tower:

✓ Perform an interpretive dance.
✓ Record yourself making a weird sound, every day.
✓ Write a love poem in Latin.
✓ Carve a duck decoy from wood.
✓ Drop a rock from 100 feet up in the sky into a tin can below.

We laughed and joked and ordered another round of amber pints. Outside, thunderclouds rolled in the darkening sky and we watched as lightning speared and splintered and forked across the bruised horizon. It was only early May, and lightning season had already begun.

My pulse quickened. Get me out there, I thought.

My body was ready, charged, for another season. I was prepared to catapult myself back into the wild.

PART FOUR

HOLDOVER

. . . the wild psyche can endure exile. It makes us yearn that much more to free our own true nature and causes us to long for a culture to match. Even this yearning, this longing makes a person go on. It makes a woman go on looking . . .

—DR. CLARISSA PINKOLA ESTÉS, *Women Who Run With The Wolves: Myths and Stories of the Wild Woman Archetype*

CHAPTER ELEVEN

When you know you are going away—to Antarctica, Mars, a fire tower—you want to make every second count before leaving. You're already nostalgic for what you'll lose: a warm embrace with a friend, choice on a menu, soaking in a hot bath, and the freedom to move wherever you want.

"Why are you going back?" a friend from Edmonton asked me over the phone. "Honestly, there's always a place for you here. Come back to the city."

But the fire tower haunted me. I craved the borderless world out there, the expansive views of black spruce and sky. I missed the place where time slows down to make space for deep thought and meditation. I wondered what the animals were doing. The grizzly, snug in his den. The migratory birds, contemplating the journey back north. I dreamt of the wildflowers, long dead, insulated beneath snow, and the safety of my yellow cabin. I thought of the long days in the cupola and even missed those, but mostly the opportunity for uninterrupted observation. Even the job itself seemed to call me back. I hadn't yet spotted and reported a wildfire, and I was determined to do so. Last season I'd been afraid of the fire tower, imagining all of the ways I would be at risk, but this year I felt the opposite. I was no longer afraid. I knew anything could happen out there, that any*one*—grizzly, black bear, cougar, wolf—could come charging out of the bush, but I was starting to remember my strength again.

"I can't put words to it," I said to my concerned friend. "I just know that I need to go back."

I wasn't going to the fire tower to hide myself away from the world, or to prove to myself that I could do it; now I *knew* I could withstand the season of isolation. I was going back, simply, because I wanted to. This year wasn't about anyone else. It would be about me. It was my choice, even if it didn't make sense to my friends and family.

"Well, don't go crazy out there," she said gently.

"Oh," I played along, "it's far too late for that."

I was trading the Heart River and its world of deciduous young-growth forest for the tower and coniferous old-growth forest. So badly, I wanted to be the first lookout to witness a wildfire this year.

I took a deep breath, inhaling the familiar beauty of the land that had held me for eight months, and whispered my gratitude to the rickety trailer on the edge of the cliff.

I wasn't flying out to the fire tower—I was *going in*. That was the expression all the tower veterans used, and it was deadly accurate. To survive the extremity of life at the fire tower, you had to be absolutely, one hundred percent, *all in*.

Lay your cards, your hand, your life, on the table.

Open yourself up to the stark beauty of the boreal, to be challenged, shaped by the land and the four elements: rain, forest, wind, and fire. Like cones from the old black spruce, let yourself go. Release what needs to be released to the fire. Learn how to grow again. Rise up from the ashes.

———

The wildfire ranger backed his truck up to the helipad at the airport in Peace River. The truck bed was packed high with my supplies: boxes of food, clothing, books, the dog's kennel, and five-gallon water buckets. In total my luggage weighed at least three hundred pounds more than what I'd brought out in my first season. Last year I'd scrimped on material belongings, but this year I'd come prepared. A helicopter, my aerial chariot, sat gleaming on the helipad.

A crew of helitack firefighters lounged nearby at a picnic table, which sat outside a mobile trailer, their man-up shack. The crew was on five-minute getaway, meaning they had less than five minutes to pile into the helicopter and race towards a location where smoke had been detected. Helitack crews, the first responders to wildfires in the forest, assess the size and category of the fire. Was it a ground fire, surface fire, or crown fire? They direct helicopter pilots to drop large buckets of water on the flames, and coordinate ground crews to lay down hoses to douse the flames or use chainsaws to fell trees and bush, to create firebreaks—gaps in the bush to help slow or stop the wildfire's trajectory and spread. The ground crews get down on their hands and knees and crawl through ash, looking for hot spots in the muskeg. Wildland firefighters are hooked on smoke and fire, addicted to the rush of the dispatch and the thrill of corralling out of control wildfires. Last season, owing to the heavy rainfall and wet conditions, I didn't get to see much of the crews.

They all wore the standard yellow Nomex button-up shirts, huntergreen cargo pants, and steel-toed chainsaw boots. The lone female member on their crew braided her long blond hair into a single rope that hung down her back. She looked confident and strong—I admired her. A part of me wanted to *be her*. Most of the men had thick, messy beards and wore huge black shades so they appeared as masked men. The only differentiating feature was the shade of yellow of their Nomex shirts. The rookies' uniforms were clean and pure as a bowl of lemons. The seasoned firefighters wore a shade of yellow that had become sullied by smoke, soot, and sweat. Together, they formed a colour spectrum ranging from ripe banana to Dijon mustard to chartreuse to ochre to a boiled-down, dirty shade of pea soup.

The leader, a stocky fellow, looked as though he'd deliberately rolled on the charred forest floor. He wore his Nomex with a kind of swagger, as if he'd fought too many wildfires to remember. It awarded him rank and authority to train and boss around the rookies, the "pukes" as they call them, bright as daffodils.

"Can you get your guys to lend us a hand loading up Trina's gear?" the wildfire ranger asked the crew leader.

They lumbered over in their giant boots and began heaving the fifty-pound boxes and bins towards the helicopter. Holly bounded happily towards the machine, and the pilot, a woman named Tara, put Holly's kennel in the back, facing my seat. Holly jumped in without any encouragement.

"You know where we're going, hey, girl?"

"You're bringing enough stuff," commented the crew leader. I bristled at his sarcastic tone, unsure of whether he was making a joke or being serious.

He was in his mid-twenties—seasoned, confident. I knew he was a respected firefighter and talented crew leader, but what did he and his crew know about enduring a season at the fire tower?

They lived on a base with electricity, hot running water, and flushing toilets. They could lean on their colleagues at the end of a hard day and were able to take days off, at least once a month. Kitchen staff prepared breakfast, lunch, and dinner for the firefighters and they even occasionally had food delivered to them while working on the fire line. But who would cook for me, or send up a hot meal at 8 p.m. when a lightning storm overhead was holding me captive in the tower? Unlike the fire crew, I would be alone with the burden of my job and the glory of my view.

"We'll make everything fit, hon, don't worry," Tara said, expertly packing up the helicopter, rearranging and assembling the boxes and bags like Tetris blocks while the crew leader and I watched. Then she stopped and placed her palm over one of the front pockets of my backpack.

"Something is wet here," she said.

My water bottle, I thought. The lid probably came loose as we were loading and unloading my belongings.

"It's wet?" I asked, coming closer to Tara.

"No," Tara said, louder now, so both myself and the crew leader could hear. "Something is *moving* in here."

Both the crew leader and I stepped closer to the moving backpack. We heard a slight buzzing noise coming from inside the pocket.

And then, horrified, I realized.

Tara realized too, and instead of waiting for an explanation, she casually walked away, busying her hands with another task, leaving the crew leader gawking at me and the backpack curiously.

My cheeks flushed bright pink.

Why was he just *standing* there?

What could I say? "Oh, my electric toothbrush!" or, "My razor must've gone off!"

The crew leader towered over me, watching like a hawk, as I discreetly reached a hand into the pocket and turned off the device.

"Well," I said wryly, looking up into his confused face, "it's going to be a long summer."

The words had tumbled out of my mouth before I could stop them. It was his turn to blush. Without a word, he walked away. Laughing silently to myself, I buckled myself into the back seat. Tara fired up the helicopter—and then, just like that, we were airborne, defying gravity, ascending straight up, up, up. The grass rolled like a wave below. The firefighters shrank to the size of yellow minions. Goodbye, men, I thought.

My last mortifying brush with civilization branded into my mind.

It's going to be a long summer.

From the front seat of the helicopter, the ranger pointed out an old burn below where wildfire had chewed up a stand of black spruce. Up above, we could see how the fire curved and meandered and jumped around several islands of deciduous trees, appearing like a misshapen mosaic on the landscape. Wildfire doesn't raze forests to ashes; islands of green aspen and unburnt pine sprang from the old burn, stands that the fire didn't touch. Regeneration was obvious, even flying at 3,800-feet elevation. Growing at the foot of the charred, burnt spruce, deciduous vegetation, grasses and willows and wild berries, had already taken root. Forest succession, or what firefighters refer to as "re-gen," was well under way.

My eye was drawn to the pockets of coral-red trees scattered throughout the forest.

"What's wrong with those trees?" I asked.

"Mountain pine beetle," crackled the ranger's voice through the headset.

Dendroctonus ponderosae, mountain pine beetle, an insect no bigger than a grain of rice. Pine beetles burrow into the bark of pine trees— lodgepole, jack, ponderosa—secreting a blue-stain fungus beneath the bark that prevents the flow of sap. The fungus girdles the tree's ability to suck up water and nutrients from the root system, which results in a staggeringly quick death. The dead pine lose their needles and turn the shade of rusted metal. In a few years they'd lose their limbs entirely and turn into a standing graveyard of what wildfire scientists call "dead wood."

Pine beetle outbreaks that occurred in the 1990s and 2000s have destroyed over 25 million hectares of pine forest across western Canada and the United States. Nearly 20 million of those hectares are housed in British Columbia. Since the 1990s, the pine beetle has attacked 50 percent of the total volume of commercial lodgepole pine in B.C. In 2006, there was a massive movement of beetles over the Rocky Mountains that spread into northern Alberta, infesting thousands of trees. Another major influx occurred in 2009, with pilots reporting clouds of beetles, while heaps of dead beetles washed ashore on lakes. It was estimated that over three million trees in northern Alberta were killed. Forest managers in Alberta responded by logging dead stands of pine beetle–infected trees and doing controlled burns to prevent the spread.

Historically, pine beetle populations were kept in check by extremely cold winters and by wildfires burning naturally in the boreal. The beetle once regulated old-growth forest, targeting only mature, dying trees. But human influence in the boreal had created the conditions for the pine beetle population to skyrocket—essentially, an expanse of old forest that needed to be cleansed by fire—and now beetles feasted on the young trees too. Plenty of scientific evidence pointed to where the blame lay: nearly a century of suppressing the natural course of wildfire, planting monocrop stands of fast-growing pine, and a rapidly changing climate that favoured the beetle's rate of survival.

Some research, although not entirely conclusive, suggests that wildfires in the northern boreal could potentially spread faster and burn hotter through forests infected by pine beetle—posing additional challenges for wildfire managers. One study in northern B.C. indicated that fire moved 2.5 times faster in beetle-killed stands than in healthy trees. Large-scale wildfires like the Horse River Fire could potentially feed on diseased forests of dead, dry wood and needles.

The Horse River Fire was no longer actively burning out of control, but it wasn't yet officially extinguished. Firefighting crews in Fort McMurray were monitoring "the black," where the fire had burned, an area that now measured over six thousand square kilometres and had spread across the Albertan border into Saskatchewan. They were using heat detectors from helicopters to fly over the charred forest and search for any remaining hot spots. The Horse River Fire wouldn't pose a risk to communities this fire season, but the Government of Alberta wasn't chancing it: crews in Fort McMurray would patrol the burnt area carefully, all summer, to ensure it was under control. The Horse River Fire had destroyed more than 2,400 buildings in Fort McMurray and the surrounding area, causing $10 billion worth of damages.

No one wanted another Fort McMurray disaster, but it felt inevitable. It wasn't a question of *if* another fire catastrophe would jump out of control, but rather *when*.

And, more importantly—*where?*

———

I felt a shot of maternal love as the fire tower and yellow cabin came into view. The dead grass, the fallow garden beds, the cabin painted the colour of pea soup—it was ugly, an eyesore even, but now it was mine. The land looked the way anyone would look after surviving six months of harsh, subarctic temperatures, of being captive to winter's teeth: weary, drained, fatigued.

Parts of me felt that way too.

My eye traced the perimeter of the yard, the length of the nearby airstrip, and I was flooded with memories of last season. *There*, I thought, that is where I walked through the wildflowers every day. *There*, that is where I stumbled upon the grazing grizzly bear. *There*, that is the place where Holly dreamt and danced beneath the ladder of the tower. Tara circled the fire tower, preparing for landing. I peered inside the red-and-white cupola, seeing the canvas cloth draped overtop of the Fire Finder. *There*, my nest, my prison cell, my cathedral in the sky.

I knew the clearing in the bush like I knew my own skin. I looked down and saw my reflection staring back up at me.

We descended gently to the world that was waking up below.

Touch down.

Home.

After we unloaded the helicopter and unsheathed the cabin, I stood on the front porch and watched them blast off from the mane of bleached-blond grass. Tara turned the machine south, and I waved goodbye as the machine faded away and out of sight.

Thump, thump, thump.

Holly beat her bushy tail against the wooden porch.

I went into the kitchen and unpacked my ingredients: flour, yeast, oil, salt, and sugar.

By heart, I measured out three cups of flour, a tablespoon of yeast, a cup of warm water, teaspoon of salt, tablespoon of sugar, and a splash of olive oil into a silver mixing bowl and worked the ingredients together with a spatula, and then my hands. I heard a faint, familiar drumming sound echoing from the forest: a male ruffed grouse, rapidly beating his wings against his chest, calling all of the females around him. If only it were so fucking easy, I mused. The elation I felt at being alone again butted up against a familiar melancholy.

Maybe the forest would catch fire and rage over my tower tomorrow. Perhaps I'd be stalked by a cougar, or a bear. Nothing felt predictable, and I knew it never would.

There was only the feeling of my hands sunk in warm, wet dough, and the knowing that when you mixed together the right ratios of flour, yeast, oil, salt, and sugar, and when the oven was hot enough to heat the cabin, the bread would rise.

Familiar voices filtered in over the two-way radio and the phone. The warm, husky voice of one of my favourite dispatchers, Dar, as I delivered my a.m. weather report. She was an experienced dispatcher, with over twenty-five fire seasons under her belt, relaying critical information to crews and aircraft. Her steady tone always made me feel safe. She used my name every so often, instead of calling me 567. Once or twice she'd even crooned, "Sweetie." I would be Dar's sweetie. I would be *anyone's* sweetie after enough time alone.

Sam had just started his third season as a lookout and his second at the same far-flung tower in the northwest corner of Alberta. His tower opened several days before mine. When he called, his voice was bright, energetic, and he laughed a gigantic, hearty laugh. A colony of rabbits were back at his tower, he said. They snuck into his garden and ate his kale and lettuce and chard, and the year before he'd watched a lynx stalk and nab a rabbit right outside his bedroom window. We vowed to talk every week.

The phone rang again and I knew who it was before even answering.

"Howdy, neighbour," crowed Ralph. "Welcome back to the forest."

Their voices were as familiar to me as faces, even though I'd never met some of them in person, and those that I had, I found I often forgot what they looked like. But what did it matter? Knowing that they were governed by the same strange factors of weather, wildfire, and isolation created a powerful intimacy, a sense of community, that was incomparable to any other I'd known. We were more than just colleagues. It comforted me to know Sam and the others were dealing with the same challenges I was, navigating emotional highs and lows.

"Lookouts are like tree swallows," Ralph once told me, referring to the sapphire-coloured songbirds that flit in acrobatic circles around our towers in May and June.

Tree swallows are migratory birds that travel long distances, wintering as far south as Panama and the West Indies. In the winter they can move in large flocks of thousands of birds. But come summer they migrate back to the boreal forest and disperse, nesting alone with their mates, raising their young. Last summer, one had darted suddenly into the cupola, hovering only inches from my unbelieving eyes. I held my breath and it flew off before I could even exhale, dazzling azure in the sun.

"In winter, we all fly away to different places," said Ralph. "But after the snow melts, we migrate back home to our nests in the forest."

The red harness fell over my shoulders like an old, worn sweater. I put on my wrangling gloves and clipped into the fall arrest system.

Climbing was as easy as breathing, but I was badly out of shape, despite all the hiking and cross-country skiing I'd done over the winter. My heart worked hard to keep up with my body. As I ascended above the trees, fifty feet up, I looked down and the old fear seized me. I began to climb using the three-point method: left hand up, right hand up, left leg up, right leg up. Rung by rung, on repeat. When I pushed through the cupola hatch, I felt both relieved and embarrassed by the effort. How many times would I have to climb to get rid of the fear of falling?

The octagonal dome was even smaller than I'd remembered, but everything was just as I'd left it on the desk: the Osborne Fire Finder, telephone, two-way radio, binoculars, sticky notes, and the bird identification book. The cross-shot map, bleached white from sun exposure; the Tibetan prayer flags hung over the west-facing window.

I rolled down the windows and the wind, cold and sharp, rushed through the dusty chamber. My eyes went galloping out the window, over the serrated back of the forest, towards the faraway horizon.

How many hours had my eyes scoured the land for wildfires? And I hadn't called in a single smoke.

Friends told me, before I left, "I hope you don't get many fires this summer." But I was eager to see smoke. Of course, I didn't want the kind of raging firestorm that would cause harm to people or property, but I wanted to see the forest lit up by lightning, a torched treetop, a ribbon of blue smoke in the distance. I was hungry to witness the natural substance of release, curling, drifting above the treeline. Mostly, I wanted to test my gut and affirm my judgment. With the first season under my belt, I knew I could do better.

Sam was baffled by my obsession with smoke. "I'd prefer for nothing to happen," he said. He strung up a hammock in the cupola and spent most of his days reading and writing and listening to music. Some days he danced up there in the sky. Wildfires tended to ignite frequently and burn vastly in the Far North, where there wasn't a soul around for hundreds of kilometres, except Sam, to play witness. He learned to feel the weather and watch for lightning, and had grown familiar with detecting and reporting smokes. Wildfires didn't get his heart rate up anymore. Sam was more of a sky monk than I was.

I wanted to trust my judgment. To tap into the singularity of purpose: to see, observe, and witness, even from afar, what the boreal forest had been doing for centuries. Ignite, smoulder, burn, and be born anew.

CHAPTER TWELVE

Lightning season was back in the boreal, bumping all of the lookouts in the Peace Country up to high fire hazard. Yesterday, I'd seen a sliver of light spearing the earth over the far southern horizon. Today's forecast included a moderate probability of lightning, which meant a high probability of wildfire. Anticipation crackled in my bones.

Around noon, a bright-red helicopter burst into view. The night before, the duty officer had called to inform me that he'd be sending a helitack crew to man up at my tower. On high and extreme fire hazard days, several man-up crews were usually dispatched to man-up locations spread out across the forest to ensure coverage of the region.

The helicopter, an AStar, a smaller machine that held a crew of four passengers plus a pilot, zipped around my tower. I waved hello to the crew of yellow men as the helicopter prepared to land on the helipad beside my cabin. These guys were going to be camped out right below my tower, all day long.

Eight days had passed without my coming face to face with *them*, people on the outside. Since I arrived, my rain barrels had remained bone-dry, and I hadn't been able to take a proper shower in over a week. What would I have to say to these men? I didn't want to answer their questions: Where are you from? Why are you here? Aren't you lonely? Bored? What are you doing after the fire season? I didn't have easy answers to any of these questions.

Letting my guard down could be excruciating. Connecting with people required exposing myself, and then refortifying after they left to endure another three months of isolation. A strange social balancing act took place out here. Every day, I woke up and reminded myself that I didn't need anyone but myself to survive the length of the fire season—that I could survive any element Mother Nature threw my way. Even when I knew that wasn't entirely true. I needed love like water.

Regardless, I couldn't hide up in the tower. It was midday and I had to climb down to check the temperatures and wind speed for my p.m. weather report. Interaction with the strange bearded creatures in yellow was inevitable. Holly, on the other hand, didn't hesitate to approach them, rocketing over to the helicopter as soon as it landed. The crew leader, a burly man named Eric, cracked open the front door. She leapt up onto the skids and he scratched behind her ears. She was always so ready, so eager, to love.

I descended the ladder.

"Hey there," called Eric. He'd rolled up his yellow Nomex shirt to reveal his tattooed forearms: an axe and the skull of an animal, maybe a bear.

"Welcome to the fire tower," I said, instantly regretting my words. *Welcome to the fire tower.* As if my tower were some exotic, five-star tourist destination—"offering the world's most exquisite views of black spruce and muskeg!"

The crew members and the pilot, a fresh-faced guy in his mid-twenties, set down their backpacks, helmets, and radios near the picnic table outside my cabin. Holly pushed her snout greedily into their bags, scenting their packed lunches.

"Some guard dog you got there," said a lean, clean-shaven guy named Thomas. He laughed and got down on the ground and threw an arm around Holly.

"How's your season going so far?" asked Eric.

He'd grown up in a small community on Prince Edward Island and had begun working wildfires when he was only eighteen years old, after starting university. Most of the firefighters were university students, often studying biology, engineering, forestry, or environmental sciences. This was Eric's fifth year fighting fires in northern Alberta, and his first year as a crew leader. Just as the lookouts migrated back, season after season, so did the firefighters.

"Pretty good, but nothing so far," I said. Nothing as in "no smokes."

"Maybe we'll get some action today," Eric said hopefully. His voice was full of optimism.

"Have you seen any wildlife out here?" asked Liam, the pilot.

It was Liam's first season flying on wildfires, and he hadn't been on a single fire yet. I imagined the pressure must be intense. Helicopter pilots were responsible for dropping off crews, then hooking up large buckets on their longlines. They dipped the buckets into nearby ponds or lakes and flew back to the wildfires, dropping loads of water strategically over the flames.

Last night, I'd spied a crab spider camouflaging itself into the yellow pistil of a wild rose. The spider had turned bright yellow, hiding, waiting to ambush a fly or a bumblebee. Crab spiders don't spin webs to capture their prey; instead, they mimic the colours of the world, wait, and attack. But I didn't prattle on to the pilot about the yellow spider, worried he'd think I'd already come unglued. I told him instead about the grizzly bear I'd seen last summer.

"You were on the ground?" he asked, impressed.

I was surprised to relax so easily into the crew's company. I realized they weren't interested in my backstory, why I'd wound up at the fire tower, or where I was heading after the season. Maybe they didn't have answers to their own stories either. Instead, they were hard-wired to the moment, knowing that one minute they could be telling a joke, reading a book, or napping in the shade and the next they'd be dispatched to a fire burning out of control. Today they were on a five-minute getaway: they'd have exactly five minutes from the moment the dispatch came over the radio to fire up the helicopter and take off towards the potential fire.

"So, you have to stay up there all day?" asked Thomas. "What do you do if . . ."

"If I have to go to the bathroom?"

"Yeah," he said, laughing.

Then I heard myself confessing to the guys about Shit Bucket, my portable composting toilet—a five-gallon bucket stuffed with dried grass and leaves—that I pulled up and down from the cupola on a rope pulley system. It wasn't glamorous, but at least I wouldn't miss a smoke because of a bathroom break.

The whole social interaction with the crew lasted only fifteen minutes before I had to get back up to the sky. The midday heat provided peak conditions for wildfires to ignite, and I didn't want to risk missing a smoke. I rushed through my weather calculations, stuffed food and water rations into my backpack, and sprang back up the ladder. I'd replay every minute of the conversation—what was said, what was left unsaid, the gestures, the laughs, the mannerisms—for the rest of the afternoon.

"Catch a smoke for us!" yelled Eric, and for a split second I wanted to hug the sweet lumberjack man. Don't be weird, I scolded myself.

"I'll try!" I answered with a laugh, already fifteen feet off the ground, climbing against the pull of gravity.

The crew came back the following day, even though no wildfires had yet ignited, and the day after that. Again, nothing happened. No smokes. No fires. They came day after day until I lost track. I grew comfortable with Eric and his crew beneath my tower, and baked them a batch of peanut butter cookies one night, which they gobbled up gratefully. A ritual was born. I told myself I was hosting my best friends.

The men read novels and lounged in the shade. They kicked off their huge chainsaw boots and practised handstands on the lawn. Holly moved from firefighter to firefighter, eager for love and attention. She curled up under Eric's arm in the shade and my heart swooned. No one was more excited than Holly on these man-up days. She was the first to hear the helicopter approaching in the distance, and would sit up, ears and eyes alert, trained above the black spruce where they'd eventually appear. I had to yell down, from one hundred feet above the earth, for her to "STAY!" until the machine had landed and the pilot shut down. The tail rotor spun dangerously low to the ground and an excitable pup could easily get in the way. Surprisingly, most days, she listened to my voice. She sat and waited, her body vibrating with excitement, tail wagging back and forth. And finally, when it was safe for her to approach the machine, I'd holler down, "Okay, girl! You can go!" and off she'd jet to the helicopter, knowing what it meant:

PEOPLE! her body language seemed to exclaim with pleasure.

Thomas took it upon himself to try to teach Holly how to fetch and shake a paw. He'd throw a stick, but Holly would only wag her tail and look at him curiously. Are you gonna go get it, Human? her comical expression seemed to ask. Eventually, he was successful with the paw trick. "Good girl!" I heard him exclaim, and glanced down to see him vigorously patting her head and offering her a treat.

One of the firefighters led a round of yoga. The bearded giants got down on their hands and knees, arching their backs like cats, then lifted their hips, coming into Downward Dog pose. I pretended not to watch, but I couldn't take my eyes off them. They were, by far, the most interesting part of these long days on high. No lightning. No fires.

The sky was a wide-open canvas of blue. The temperatures climbed high into the twenties, but the clouds were docile beasts, seemingly uninterested in swelling up into storm clouds. The eleven-hour days stretched long and hot, one after another, every day a clone of the last.

"Do something," I whispered to harmless tufts of white cumulus floating by my window.

"See anything?!" Eric hollered up at me from the ground. I craned my neck out the window.

"Nothing, sorry!" I yelled down.

They flew away at the same time every day, minutes after 7 p.m. One day I waved casually as they departed, assuming they'd probably be back tomorrow, and the day after that. They'd be here until lightning split the sky and speared the forest below, until I cried, "Smoke!"

An hour later, I gathered my lunch and Thermos, packed up my bag, stepped into my harness, and rolled up the windows. I clipped my composting toilet onto the pulley system and began lowering the bucket one hundred feet below. Something caught my eye to the north—was it dust?—and as I leaned over for my binoculars, I lost tension in the rope. The bucket, heavy with urine and shit, took off, careening violently towards the ground. The rope burned through my fingers and I had to let go, horrified, watching Shit Bucket hurtle

down,

down,

down

"Noooooooooooooooooooo!"

Holly looked up, the white bucket plunging down towards her.

"Holly, GET OUTTA THE WAAAAAAAAAAAY!"

She leapt sideways a second before impact.

SPLAT!

I gaped down at the crime scene. I was speechless. Had I just dropped Shit Bucket from one hundred feet up? I began to laugh loudly, hysterically, and climbed down the ladder, laughing so hard I was crying. I felt as though I was becoming a character in a comic strip.

THE ADVENTURES OF TOWER GIRL AND TOWER DOG!

I could see myself—my alter ego—a hairy-legged, wild-haired Tower Girl with a cape around her neck. She stands up in the cupola, her eyes bugging out of her head, watching Shit Bucket, in slow motion, career down to earth and go off like a bomb. CABOOM! Tower Dog, her faithful sidekick, jumps out of the way, right into the limbs of a tree.

SHIT HAPPENS, the caption would read.

Tower Girl doesn't give a shit about Shit Bucket. She laughs. She tosses her head back and throws her laughter onto the wind, waking up the wild things that lurk beyond the spruce. Tower Girl stops midway down the ladder, dangling dangerously between the clouds and earth, and lets out a wolf call that reverberates throughout her Queendom, so the whole forest can hear her joy.

Holly eyed me curiously, probably wondering, Has the Human totally lost it?

Maybe I had. But I was also struck by the realization that it felt good to howl, to give in to a belly-deep, genuine laugh, and to not give a shit—about Shit Bucket, or missing a smoke, or the mistakes I'd made— or anything beyond my home in the woods.

By eleven o'clock in the morning, the molten sun poured out of the sky and the humidity soared. The cupola felt like a dry sauna, so I cracked open all the windows, but the limp wind barely moved the anemometer. I would've described the wind as calm in a weather report, but that wasn't quite accurate. This kind of weather was a recipe for disaster, for wildfire. My neighbour, a seasoned lookout to the east, sent me a text: "Trina, stay vigilant today. There will be fire before the day is over."

I looked southwest and spied the parade, a procession of cumulus approaching from about forty kilometres away. I wondered how long it would take to wake up these sleeping clouds. Another hour? Two hours? I was grateful when I heard the sound of the helicopter approaching.

"Something has got to give today," I said to the crew, speeding through my weather calculations before rushing back up the ladder. No banter today. I was all business, ready to catch my first smoke. I climbed up into the cupola just in time to catch the first strike from a cell to my southwest. Lightning shot out of the bruised belly of a huge cumulonimbus cloud.

"First strike!" I yelled down to the guys, before radioing it in.

"Wahoo!" Eric cheered from down below.

One of the crew members sat up, suddenly awake, and laced up his boots. They began to gather their belongings, anticipating a dispatch, but the radios stayed quiet. I waited another five, ten minutes before spotting a second strike. Nothing but moisture rose up out of the forest. My eye lingered on a few of the spooks. Was it a smoke? *Yes?* No. Definitely not a smoke. I put the radio mic down.

"See anything?!" Liam, the pilot, hollered up. He paced back and forth like a caged tiger, eager to be up in the air, flying, and bucketing on his first wildfire.

"Not yet! I'll let you know!"

I was revelling in the camaraderie of the crew, their reliance on my best judgment. I was the only one who could see the forest and cry smoke, and it made me feel, even for a brief second, a part of their team. An honorary member. Tower Girl and her yellow posse.

The wall of black pushed northeast, attracting more clouds and gathering strength. My neighbour to the east called in a dry strike from the same cell. Aha, *there*. I spied a white filament, ahead of the storm, spearing the coniferous tops. The storm was pushing directly overtop of his tower. "I've got hail," he said over the radio. "Alpha one-zero," which described pea-sized hail. Alpha one-one was for grape-sized hail, and alpha one-two described walnut-sized ice chunks.

My phone rang and I jumped at the sound.

"Hello?"

"Hi, Trina, this is the duty officer calling," said a male voice. "Can you tell Eric to get the guys off the ground and head out on a patrol?"

I shouted down to the crew and waved goodbye as they trotted back towards the helicopter, loaded their gear, and flew off to patrol the areas of forest to my northeast for wildfires. I worried they'd catch a smoke from the air—"steal a smoke," as some lookouts were prone to saying. No lookout wanted to miss a smoke. And who could blame us? We spent hours, days, weeks, months, and, for some, multiple seasons scanning the horizon for wildfires. Aircraft patrols often had the advantage over lookouts, flying over low-lying smokes that would be difficult to see from the fire tower. No one liked to be "scooped" by patrols, even if we were all on the same team.

"Don't steal my glory, guys," I said to the empty space they left behind.

The red helicopter shrank into a fleck of red paint against a wall of black. I continued to watch, but nothing happened—no strikes, no smokes, no fires.

Maybe not today, I thought glumly.

I poured myself a cup of coffee and turned on a podcast, CBC's *Vinyl Café*, then sat down in the swivel office chair and kicked my boots up onto the window ledge. I leaned back as host Stuart McLean's familiar voice filled the cupola.

The current episode was about Dave's daughter, Stephanie, and her tree-planting adventures in northern Ontario. It was one of my favourites.

"Stephanie landed a job as a tree planter this summer. She thought it was going to be a *ton* of fun," Stuart McLean said wryly. He paused and the audience laughed. "She heard you could make a load of money in no time flat, planting trees. So she signed up and found herself on an old school bus, heading north out of Thunder Bay, in the early days of May—full of hope."

McLean described Stephanie's first day on the job. Her boss dropped her off at a scraggly patch of alder-choked land with a box of 2,500 white pine seedlings. "Make it happen, man," the boss said to Stephanie. By the end of the day she'd only planted 500 trees. Impossible, she thought, 2,500 trees would be impossible. Every day was a blur, the same blur—up at 6 a.m., down to the bus, and out to the cutblocks to plant. The weather wreaked havoc. The wind and the rain lashed the planters.

"When stabs of lightning began to jump out of the sky," said McLean, "Stephanie was so determined to push her numbers up that she just kept on planting. And then it started to rain. It rained for a solid week. And then the blackflies came. Swarms of them. Like someone was pouring them out of a giant jar."

Stephanie's hands cracked and bled. Her body ached. But what bugged her the most was not reaching the goal of 2,500 trees. Planting trees was "such a simple thing" and yet 2,500 seemed out of reach. She agonized over her technique, her tools, her sleep, trying to get her numbers up. She cornered Rob, one of the seasoned planters, and demanded from him, "Tell me the secret."

And then I looked up and nearly spilt hot coffee all over myself.

g
n
i
s
i
r
e
k
o
m
s

It was undeniable, a thin scribble of dark grey smoke ribboning up over the northeastern ridge, halfway between my neighbour's tower and my own. My mouth opened, a perfect *Oh* exhaled from my lungs. *Just like that.* A fire was born in the boreal and I was the first to see it. I swung the Fire Finder around, lined up the black smudge through the crosshairs, and scribbled down the bearing onto the pre-smoke form, estimating it at roughly thirty kilometres away.

Distracted, I forgot to turn off the podcast. Stuart McLean's voice droned on in the background: "The secret to planting more trees is to plant more trees," he said.

I picked up the mic and shakily delivered my smoke message.

"XMA26, this is XMA567 with a pre-smoke."

"XMA567, go ahead," said the dispatcher.

"I've got a smoke at 55 degrees, 20 minutes. It's grey and coming straight up."

"That's all copied," she said.

Did I detect a note of happiness in her voice?

My neighbour to the east jumped on the radio with a cross-bearing. "Trina's smoke is coming up black!" he told the dispatchers. "It must be burning in coniferous."

Only seconds later, Eric's crew was racing towards the smoke.

As they circled the darkening plume, I heard him say "confirmed wildfire" over the radio. Now the smoke was theirs to manage. The drama of battling the blaze, suppressing the forest's old, evolutionary instinct to burn, the gruelling physical work, the responsibility—it all belonged to them. But for a moment, a fraction of a second, the smoke was mine and mine alone. A black, curling apparition that only I could see.

Stuart McLean preached on.

"When it's time to plant, plant. When it's time to eat, eat. Whatever you're doing, just do it," he intoned, his voice filling up the cupola. "You can worry about the rain all day, or you can plant and make money. The rain will pass. As for tomorrow, tomorrow doesn't even enter into it." Stephanie, said McLean, finally broke 2,000. And then she hit 2,500—despite the blackflies.

I stayed up in my tower, watching the smoke persist, twisting charcoal black against a magenta sky. The seasoned lookout to my south called on the phone to congratulate me. "You'll never forget your first," she said fondly, and I believed her. I watched until the smoke faded into the horizon, until the ash settled on the shoulders of the firefighters, and the old moon gazed down upon the forest.

CHAPTER THIRTEEN

Rainy days brought relief to the lookouts. We called them Romeo Whiskey days, which was lookout code for rain showers. The clouds coalesced, rolling low overhead and providing a generous shower that doused smouldering hot spots in the forest. The anxiety over missing a smoke washed away with the persistent rain, as did the calculated worry of rationing water. As soon as I heard the sound of water pooling at the bottom of the rain barrels, I collected a big pot to boil on the stovetop for my first real shower in nearly twenty-one days.

Holly was happy to have me back on the earth. She followed me everywhere. Without my even looking down, my hand reached for her soft ear. We seemed to move as one body.

I relaxed into the rhythm of rain on the tin roof, puttering from room to room, woollen socks pulled up high to my knees. I boiled water for coffee, curled up in the old La-Z-Boy chair, and called Sam's tower, hoping the overcast skies had roamed to the Far North.

"What's your hazard today?" I asked him.

"Low," he sighed. "Thank god."

He was also boiling water, for a bath. He'd bought a gigantic plastic bin that would serve as both cargo and a bathtub. Our rainy day rituals weren't so different: hot tea, reading, writing, napping, and taking the time to cook, or bake a loaf of bread. Lookouts used low hazard days to get caught up on a never-ending list of neglected chores: cleaning the cabin and engine shed, doing paperwork and laundry, and, when the rains finally let up, mowing, weed whacking under the bear fence, and yardwork.

Ralph called and jokingly asked, "Should I send over a canoe to rescue you?"

Rainy days were also for dreaming about planting gardens. I'd already sowed my root crops—Alaskan potatoes, tricoloured carrots, and yellow onions—but held off on planting the peas and squash and beans and herbs. "Don't plant until after the May long weekend" was the rule to abide by in northern Alberta, where frost and snow was always possible in late May and even early June.

I dug a box of vegetable and flower seeds out of storage and sorted them at the kitchen table, pencilling down my ideas onto the back of a pre-smoke form. Peas here, sunflowers there. Where to plant the beans?

I thought of Akello at our low wooden kitchen table doing the same thing: rooting through a bag of seeds, dreaming of planting. It was late May and he could plant one more cycle of an early riser, such as radish or Swiss chard, before the dry season in southwestern Uganda. After June, there was no point in planting anything; the rains ceased and sun scorched the land for most of July, August, and September. Before I left, we'd spoken of planting cassava, a heat-tolerant crop that grew twelve to fourteen months before harvesting. Cassava was the crop that defined Akello's childhood.

We still wrote to one another every few weeks, focusing only on the logistics of our lives. It was too painful to go any deeper, but even the smallest details of our routines were loaded with symbolism and memory.

"I harvested potatoes and basil from the garden today," he wrote.

The Italian herb I'd smuggled over in my bags flourished under the

Ugandan sun. Akello had grown to love bruschetta and basil tomato soup. He let several of the plants go to seed and stored the seeds in small paper envelopes.

Our love for one another had changed us both; it had even changed our taste buds.

I eyed the small bag of corn seeds and scribbled "corn" in the same garden bed where the beans would grow. It didn't really make sense to grow corn at the fire tower. It was unlikely to thrive at this northern latitude, even with the long summer hours of daylight, because the muskeg soils were slightly acidic—not ideal for corn. Also, corn was a thirsty crop. I wasn't sure what kind of fire season it was going to be— hot, dry, wet, or cool. It seemed risky to plant such a demanding crop, but I felt loyal to the ritual and memory of planting corn with Akello. It was the first thing we'd ever done together, as friends. I loved the way corn emerged from the earth—a rolled shoot of bright green that always stoked my sense of hope for the growing season. As the corn plants grew thicker, taller, they supported the life of other crops—climbing beans that wound round and round their sturdy stalks—and their leaves fluttered in the breeze like party streamers.

A few days later, I planted corn in my garden the way Akello taught me, barefoot. I used the hoe to make a divot in the soil, dropped a couple of seeds, then used my foot to kick soil over them. Repeat.

As I planted, my heart felt free and happy. It was a way of connecting threads and storylines, of making sense of the *here* and *there*. A homage to our friendship and story.

Holly and I went walking through the stand of black spruce and mossy swamp, out into the grassy opening where the fireweed was poised on the brink of blossoming bright fuchsia. Dew adorned the edges of everything, grass and wildflowers and leaves. The forest sparkled and sang with birdsong. I breathed in petrichor, the smell of rain after a long dry spell.

A pair of Australian scientists came up with the word *petrichor* in 1964 when they found that, during a drought, some plants excrete oils. Soil bacteria emit oils too. When rain falls to the earth, water mixes with plants and soils, and these oils—like essential oils—blossom in the air. Petrichor: a sweet, evocative smell that intoxicates the senses. It led me down a trail towards the airstrip.

And then I saw something I'd never seen before. My eye, drawn to the unfamiliar, to a tiny altered detail on a landscape I observed day after day after day. It was the mark of a human.

A bouquet of flowers.

Well, a bouquet of pussy willows. Someone had wedged the bouquet into the door handle of the old tin bunker that had been built decades ago but was no longer used by firefighters. Wild grasses and flowers grew tall around the edges of the empty shack. There was a long claw mark along the door, probably from a curious bear.

The soft willows were tied at the stem with a thin strip of black electrical tape. I felt a hard object taped to the stem, and I unravelled the tape to discover a tiny white stone—smooth quartz, no bigger than a hummingbird's egg. I rolled the stone between my fingers. Almost instinctively, I pressed the soft willows to my face, tracing my lips with the velvet buds. They reminded me of being a girl.

A smile blossomed on my lips.

Yesterday, a lone pilot had landed and dropped off half a dozen barrels of jet fuel. But I suspected he had only been the messenger. Who would send such a sweet offering?

I floated back to the cabin, delighted by the surprise. Yet I scolded and cautioned myself, Don't be so romantic. It was just a bunch of sticks tied together with tape—not food, or water, or anything practical to help me survive alone in the woods. But after nearly six weeks of isolation, the bouquet of pussy willows gave me the feeling that someone out there was thinking about me; that, as invisible as I felt—erased by distance, concealed by the cover of forest—I hadn't been forgotten.

———

I thought often of my grandfather John, who'd passed away two years earlier, soon after my return to Canada. A keen observer of people and nature, he would have loved the lookout life.

He had worked as an air gunner during World War II, operating the radio and one of the machine gun turrets, flying through the UK, Iceland, India, and Egypt. In the UK he flew in Sunderland planes, or "flying boats," a model of airplane that's strikingly similar to the Ducks, a type of aircraft used in modern times by Forestry for water-bombing fires. After the war he became a lawyer, and later an English teacher and high school principal. My grandfather was also a passionate wood-carver. He felt for the heart of pieces of wood, listened to the grain, and carved out ducks and loons and walking sticks, like the one he gave me before I left for Uganda.

We visited my grandpa and nana's cottage in Kenosee Lake, Saskatchewan, nearly every summer, and he loved to share stories from his childhood with us. He immigrated with his family to the Canadian prairies from Ireland in 1926 at only three years old, so his earliest memories were on the homestead plot outside Wolseley, Saskatchewan, growing up a farm boy. He and his older brother ran wild on the land, catching garter snakes, rescuing a baby owl that had fallen from its nest, and target shooting with a pellet gun from the roof of the barn. Horsing around, they'd accidentally fired the gun once inside the house—assuming it wasn't loaded—and blew his mother's canary into yellow oblivion.

He had been a gifted writer and wrote extensively about his experiences flying in WWII. He left Wolseley at eighteen with big ideals and a naive, boyish spirit, but flying overseas and experiencing the tragedy and toll of war first-hand hardened his idealism. He wrote a memoir about the years he spent flying in WWII, recalling a near-death experience flying through a storm over Iraq, and how he'd carved his and my nana's initials into the Great Pyramid. Near the end of the war, he'd flown right beside a German four-engine aircraft, locked eyes with the German pilot, and thought, If he doesn't fire, I won't fire. Neither of them fired a shot. "In over fifteen hundred hours of flying in the war, never once did I fire a shot in anger," he wrote.

When I embarked on my first solo trip to Central America at nineteen with a youth humanitarian project, my grandfather and I began writing one another letters. My travels reminded him of his journeys during WWII. When I came back to Canada, the correspondence continued, though we soon switched from letters to emails. We shared a hunger for knowledge and exploration, and a fervent idealism that could sometimes get us both in trouble.

He too was drawn to the North. In 1985, the year I was born, he was working for the provincial government, with Municipal Affairs, in La Ronge, a remote First Nations community in northern Saskatchewan. He saw and understood the disconnect between northern communities, particularly First Nations communities, and southern bureaucrats, how decision makers in the South often controlled resources in the North. He once submitted a political cartoon to the local newspaper just before an important politician from Regina was scheduled to visit La Ronge, which infuriated his government employers. He'd drawn a not so flattering caricature of the southern bureaucrat—an obese man from the colonial era, wearing royal clothes, pompously parading by a crowd of amused northern folk.

"I was nearly fired for that one," he chuckled.

My grandfather was fascinated by ravens, the large corvids that dwelled in La Ronge. He wrote an essay entitled "A Love Affair with Ravens" in which he describes the way a group of twenty or so ravens would gather in a loose circle, croaking and clucking and crying at one another. Once, he felt the rush of a raven's feathers on his skin. It seemed to me that ravens were a symbol of both community and solitude for my grandfather.

What creative mischief he might have stirred out here in the bush. What stories he would have spun. I missed him, deeply. I'd decided to let go of my sadness and regret around the miscommunication we experienced before his death. The decades-long connection we shared meant more than a few months of silence.

I thought of him every time a lone raven swooped low around my tower, or landed in the branches of a tree below, croaking and chattering in a language much older than me.

A letter fell out of the sky today.

This is the first of as many letters as you're willing to accept.

My hands shook with pleasure as I read the handwritten letter up in the cupola. A firefighting crew, the messengers, lounged below on five-minute getaway. The letter was from the same person who had made the wild bouquet, the same man I wandered the valley with last autumn, chasing after elk. *Jay*, he signed at the bottom of the notepad paper, his writing neat and measured.

I hope you'll find some enjoyment in these letters. Maybe you'll laugh at my grammar and how the school system failed me. Or, worst case, you can always use them as fire starter.

The sound of my own laughter echoed off the cupola walls.

Jay and his crew mates were manned up at a tower over a hundred kilometres away, waiting for a dispatch to a wildfire, not unlike the crew that lounged below my own tower. He had come back to Peace River for his fourth season as a sub-leader on a unit crew, a twenty-person crew made up of five sub-crews; and like many of the other men and women who worked wildfires, he was a seasonal migrant, a transplant from southern Ontario. Some of the firefighters had notched up to ten fire seasons under their belts. They worked on the front lines of fires until they were offered ranger positions, or until they migrated back to the real world.

Like me, Jay was moored to the moment, and wrote to me about the wild things that caught his eye.

I just stumbled upon a small purple flower. My guess is that it's a northern purple violet, but it reminded me of a lady slipper, which are so pretty just like . . .

Me. ☺

And you thought I was going to say you.

We'd both travelled around the world. He had spent long months in South America and New Zealand and California and northern Canada, but he felt most at home on the water. He'd been a canoe guide in northern Ontario and was planning to paddle the Notikewin River, a tributary of the Peace River, on his days off. He was drawn to the same places—vast, open, and unpeopled—that now set my heart ablaze.

I was physically alone in the tower, while he was surrounded by the same four people on his sub-crew, every day, all day, for the whole fire season. Yet we seemed to share the same emotional geography, and maybe the same instinctual longing for exploration—and companionship. His words fanned a dormant flame in my chest. I looked south beyond the Notikewin River as though I could see him from afar, curled beneath the shade of a helicopter, writing, remembering, dreaming—and waiting for something to burn.

The tenderness of the letter disarmed me. I picked up a pen and began to write back to him, glancing up every few minutes to make sure the forest wasn't on fire. I wrote with an openness that I hadn't felt for so many seasons now. The last person I penned actual letters to was probably my grandfather. How is it that I felt more myself with a pen in my hand?

Later that evening, I dialled Jay's number.

"Hello?" he answered.

"Hey," I said tentatively. "It's Trina. I just wanted to say thank you for the letter. That was really thoughtful of you. It made my day."

"I'm happy you called," he said. "I make a point of listening to the afternoon weather reports on the radio so I can hear your voice. I've been wondering how you're doing out there."

He's kind, I told myself. He's made a consistent effort to connect with you. Even so, I was afraid of him. I was fearful of letting myself fall in love again, but it was comforting to know that Jay was out there, thinking of me, and I couldn't deny that I'd also thought of him, chasing after the elk song with me.

How close we'd come to holding hands and brushing lips.

Jay's voice was a balm against the loneliness. We laughed and traded stories back and forth and talked until the midnight sun plunged beneath the horizon of black spruce.

"Can I call you again?" he asked before saying goodbye.

"Yeah," I said softly. "I'd like that."

Melanophila acuminata, a centimetre-long beetle native to the Canadian boreal forest, is biologically drawn to wildfire in order to mate and reproduce. *Melanophila* is Latin for "lover of blackness," as the female beetles lay their eggs in the black, charred bark of fire-ravaged trees. The beetles' belly is equipped with over seventy infrared sensors called sensilla, which detect the heat radiation produced by wildfires. Miraculously, these beetles can sense a wildfire from up to eighty kilometres away. Silently they go scuttling across continents of muskeg, lured by what's burning hot and black. In the aftermath of wildfires in the boreal forest, the lovers of charred blackness are born again.

CHAPTER FOURTEEN

White light pulsed outside my bedroom window, followed by an earth-shaking clap of thunder over the cabin. Holly trembled with fear, cowering in a tight, anxious ball. I glanced at my clock: 5 a.m. I jumped out of bed and ran to the kitchen window to gawk outside at a strange, apocalyptic sky, the colour of a black violet. Lightning ripped the sky in half and stabbed at the stand of silhouetted lodgepole pine only a few hundred metres away.

I couldn't hear rain against the tin roof. These were *dry strikes*! My heart hammered in my chest.

"Stay inside your cabin," they told us at training. "Do not go out into a storm."

My whole body was a charged wire. Fuck it, I thought. I obeyed instinct: snatched my camera and hurled myself outside into the pandemonium. Holly was hot on my heels, absolutely terrified by the sky on fire, but loyal to my every move.

BOOM!

"Ahhhhhhh!" I let a blood-curdling scream rip out of my lungs as I sprinted towards the black spruce, grinning like a madwoman. I fell to my knees amongst the yarrow and purple-and-yellow star-shaped asters and white, unripe bilberries. Holly practically climbed onto my lap. Her body trembled, so I wrapped my arms around her and whispered into her ear, "It's okay, girl."

Lightning split and fissured the purple-black sky. I witnessed a strike flash up-down-up-down in the same spot like Morse code. TA-TA-TA-TA-TA!

What they say about lightning not striking the same place twice? A fucking lie!

The black clouds rolled and seethed. I tried to photograph the moment but felt a loss holding the camera to my face, reducing the view to a tiny, limited box of light. Instead, I bowed down on my knees, mesmerized by the carnage of light. When the storm finally fatigued—sixty strikes later—I stood up, my limbs trembling.

Lightning causes around 40 percent of the wildfires in Alberta every summer. Molten-hot strikes often occur in large swaths of remote forest—on essentially unpeopled landscapes—and catalyze the largest, hardest-to-control wildfires that burn furiously through the northernmost parts of the boreal. Humans may bring about the majority of wildfires every year, but lightning often causes the largest wildfires on the landscape.

The Horse River Fire of Fort McMurray wasn't the largest wildfire in recent Canadian history. In 2011, a violent lightning storm ignited a wildfire northeast of Fort McMurray, and the flames grew into what became known as the Richardson Fire, which burned through the Richardson Backcountry, 700,000 hectares—nearly two million acres—of forest. It was the largest documented wildfire in Alberta's history since 1950, but because the wildfire didn't threaten any communities or human property, it barely made the news.

Lightning has historically been Mother Nature's preferred evolutionary catalyst for sparking wildfire in the boreal forest. It's a naturally occurring phenomenon that helps the boreal forest to regenerate life, to grow anew. During peak storm season in Alberta—late May through early August—there are tens of thousands of lightning strikes on any given day. Historically speaking, this isn't new. Lightning-caused wildfires belong in the boreal.

But according to a 2014 study published in *Science*, climate change could be affecting the frequency and intensity of lightning storm activity in the boreal forest. Scientists are predicting that for every one-degree increase in global temperature, there will be a 12 percent increase in lightning activity.

Where there is lightning, there is fire.

The smoke was an apparition on the western horizon, a small white dot against an ocean of blue. It grabbed my eye, spotlit by the sun, impossible not to see. I couldn't see the trees engulfed in flames, only the release of moisture and carbon dioxide rising up into the atmosphere. The fire burned brightest where it blistered the bark, brilliant white, then faded and curled into a grey column, tugging low along the horizon. "Drifting low" was how I would describe the smoke in my pre-smoke report. I watched it for a few seconds, awestruck.

"I've got a smoke for you guys!" I hollered down to Eric and his crew. "To my northwest!"

A chorus of cheers broke out below and a sudden flurry of yellow as their long limbs reached for their books, brown bag lunches, hand-held radios, and water bottles. They shoved everything into their backpacks just as they shoved their feet back into their untied chainsaw boots.

"Good luck!" I yelled.

They made a frantic, almost comical dash to the helicopter.

"26, this is 567 with a pre-smoke," I spoke clearly into the radio mic.

Pre-smoke. I wasn't a fan of the word. It seemed to suggest not yet a smoke, perhaps just a ghost, or a cloud of dust, pollen, or a trick of the light on the lookout's eyes. It wasn't until the crew flew overhead of the detection site that it could become a legitimate wildfire in the eyes of the government.

But I knew in my gut it was real smoke, a real wildfire that was burning out of control. Cause of the fire? No doubt one of the violent dry strikes from the crazy morning storm.

"26, this is Yankee Whiskey Bravo," said Eric over the radio. "Smoke in sight. It's white and drifting low. We're approximately fifteen miles back from the smoke."

"That's copied," the dispatcher replied. "Will you be requiring tankers?"

"No tankers required, maybe just another HAC crew, if possible," said Eric.

The smoke column braided itself into a horse's tail. The fire wanted to run, but she probably wouldn't go far. Although the fire hazard was creeping up again, the earth was sopping wet, a far cry from the May conditions when the forest was a matchbox waiting to be struck. I imagined all the tiny fire-loving creatures, including the fire beetle, scuttling towards the wildfire, eager to lay their eggs in the charred bark of the burnt trees. In a matter of only weeks, fireweed and other pioneer species would germinate in the ashes of the burn. Deer, moose, elk, and bears would come to forage on the new green growth. And somewhere deep in the charred floor, the nitrogen-rich ash would fertilize the opened coniferous seeds waiting to grow again.

"Burn, girl, burn," I whispered, watching through my binoculars. But it wouldn't take long for Eric and his crew to call BH, being held, on the wildfire.

By 8 p.m., the smoke was barely discernible, a grey blemish in the distance.

———

I'd found love in different places.

Chased off the streets by torrential rain, my friends and I wound up in an art gallery in central Cuba, where we stumbled into the company of Cuban painters. I locked eyes with a slender, curly-haired man named Alián. "*Entonces*," he asked me early on in our friendship. "*Quieres ser amigos, o algo más?*" Do you want to be friends, or more? Oh, I wanted more. I was twenty-six years old. I wanted the whole universe: adventure, new perspectives, love, sex. We devoured one another and held hands only at night, when his neighbours were asleep, when no one would talk. "It used to be illegal to be seen with tourists, you know," he told me. "Cuban neighbourhoods have eyes." He loved Cuba, but he wanted to travel to Miami to work and study English. Before I left, he gave me one of his abstract paintings of the *palma*, the Cuban palm trees. The painting reminded me of the view from his bedroom window: palms bending in the distance, heat rising up off the hot concrete buildings, blurring the view.

A mirage.

I had followed a friend to England for love and once jumped off a thirty-foot cliff in the Okanagan, trying to impress a summer lover. He gave me a daisy and I told myself we were soulmates. I cancelled many flights back to Canada to linger a while longer in the arms of boyfriends. I'd learned how to say *I love you* in so many languages. My heart was wide open, a parasail. If there was wind, I'd fly. *Vámanos*, let's go, I always said.

But since last summer, I'd become fearful of love, or rather of losing love. I thought that exiling myself to a place as far-flung and improbable as a fire tower would protect me.

Jay called nearly every evening and we spoke for hours on the phone, talking and laughing and talking some more, until the stars burned brightly into constellations and the northern lights danced themselves onto the sky.

I was growing deeply fond of these late night conversations with Jay, and although we'd only ever spent a handful of days physically together,

I felt a familiar magnetic pull, a feathered rush every time he sent a text or I heard his voice on the other end of the phone.

"Do you see what I'm seeing?" he asked, incredulous, staring up into the black vacuum of space.

"Yes," I answered breathlessly, as the stars pressed down on us.

I had never fallen in love with a voice before.

The forest trembled like a butterfly, shivering on the brink of blossoming and bearing fruit. The goldenrods rocketed out of the soil and the lupines wavered in the wind, donning purple bells on their towering stems. At the edge of the clearing, the fireweed was just starting to erupt. Once the edge of the bush ignited with colour, berry season would be underway. Raspberries, dewberries, cloudberries, and bilberries would soon grow fat and dangle seductively off their stems.

The wild strawberries, no bigger than my pinky, were already ripening beneath their furry leaves. I popped a sun-warmed strawberry on my tongue: a minuscule bomb of sweetness exploded in my mouth. It tasted like a forbidden fruit, an edible jewel in the Garden of Anywhere-but-Eden.

I felt joy in my bones. I felt at home.

As the summer was unfolding, I was learning how to see rather than just look fearfully out the window. I no longer feared that I'd miss a smoke and let the forest burn down. The weather sank beneath my skin. The temperature and humidity and wind speed became more than numbers I jotted down into the a.m. and p.m. reports. My body was a barometer.

I grew bolder on the fire watch. Bold and proud. Every wildfire I detected was a badge of honour. *PWF034, PWF075, PWF089.* I joked that I'd tattoo the fire numbers on my skin.

I climbed the tower with muscled ease and knew the difference between a smoke and a spook. The forty-kilometre radius of forest was imprinted on my mind. I could pick out every alteration, every cutline, cutblock, well site, and communication tower. As I hiked through the bush, every step was grounded in the present, my senses were on high alert, ever ready to face whatever was around the corner. When I saw my reflection in the mirror, the woman looking back at me appeared stronger, healthier, and even happier than the previous summer. I allowed Jay into my heart while looking for smokes, cooking dinner, falling into bed every night. As I looked out over the forest, *love*, the four-letter word, couldn't encapsulate the feeling.

It was more like,

F-R-E-E-D-O-M

The words on the page flowed and I began to write myself anew. It occurred to me that I could be anywhere in the world and feel at home in my own skin as long as I was writing. I published a few stories about working at the fire tower for a government blog and remembered the empowering feeling of sharing my work with the world beyond. My sense of self and purpose was slowly being restored, and I began to wonder: Could I tell those stories? Maybe. One day, maybe.

Looking out from my cupola, balanced between forest and sky, I felt my confidence soaring.

But I forgot that when you think you know everything about nature, you really know nothing.

"Nothing is going to burn today," I heard myself say matter-of-factly to a helitack crew. I'd just briefly climbed down to do p.m. weather and get my lunch ready.

I hadn't met the crew before. The leader was a stocky, blue-eyed, clean-shaven guy named Grant. He nodded politely at my statement, although he might very well have asked:

"What the hell do you know, Tower Girl?"

But I'd been watching my area of forest carefully. Last week I'd observed lightning pierce the eastern ridge between my neighbour and me, but it had also been rainy and foggy for the past several days. Although we'd received an abundance of rain, we were on high hazard, most likely because of the high winds. Even so, it didn't feel like a burning day to me.

My neighbour to the east agreed with me. "Trina," he said, "it's way too wet out there."

Clearly, the duty officer thought otherwise. He'd dispatched Grant's crew to man up at my tower all day on a ten-minute getaway. After climbing down, I made small talk with Grant and his crew for several minutes, then excused myself into the cabin to get my lunch ready. As I was heating up last night's pasta over the stovetop, I heard my easterly neighbour's voice over the radio.

"XMA26, this is XMA685 with a pre-smoke," he said.

Those two words, *pre-smoke*, a fire alarm wailing over the loudspeaker. *Fuck!*

I scrambled into my boots, burst out of the cabin, and flew past the crew. They were gathering their belongings, already anticipating a dispatch to the smoke detection. I struggled into my climbing harness and sprinted to the foot of the tower. Holly gave chase, as though she could sense the urgency. I dashed up the ladder faster than I'd ever climbed before. Above the treetops, fifty feet off the ground, I glanced east towards my neighbour's tower.

My jaw dropped.

The smoke was coming up huge and black.

The helicopter was spooling up. I needed to radio in my cross-shot bearing on the smoke before the guys got off the ground. My cheeks flushed with equal parts exertion and embarrassment as I raced up the ladder.

Nothing will burn today, my cocky prediction stuck on repeat. Why had I said that?

I burst through the cupola hatch door, panting loudly. My lungs hurt from the rapid-fire exertion. After nearly a week of nothing burning, my Fire Finder had become a coffee table. A pile of books, an orange, a half-finished bottle of blue Gatorade, and a roll of toilet paper littered the compass's surface. I threw everything onto the chair and swung the compass around to line up my shot. The smoke was a black monster.

"The smoke is really taking off!" I heard my neighbour's panic-stricken voice over the radio.

Breathless, I jumped on the radio with my cross-shot. "XMA26, this is XMA567 with a cross-shot on the smoke," I rushed.

"Go ahead," said the dispatcher.

I panted off my bearing a fraction of a second before the helicopter lifted off the ground.

"Smoke in sight," said Grant, his steady voice signalling to dispatch that they had a wildfire on their hands. "We're going to need air tankers on this one."

As Grant's crew flew over the smoke, orange flames devoured the tree-tops. Swept by the winds, the fire had flourished into three hectares—six football fields! I imagined I was sitting up in the nosebleed seats of a stadium and looking down on the raging inferno. I should've felt excited to witness such a wildfire, the biggest I'd ever seen at close range from the fire tower. And I'd passed along my cross-shot before the crew lifted off the ground.

But even so, my confidence was shaken.

Nothing is going to burn.

The radio exploded with voices of firefighting crews, helicopter pilots, and the Bird Dog and Electra air tanker planes, which were already racing towards the fire. Through the binoculars I watched the mustard-yellow Electra plane release a load of fire retardant—red as a shade of Cherry Bomb lipstick—over the black, billowing smoke.

It was a holdover fire—a dormant fire, which had ignited nearly a week ago, after lightning split open the skies and shot strikes. My neighbour and I had been watching the area, and Grant and his crew had been patrolling it as well. Miraculously, the fire lay hidden, out of sight, burning deep in the muskeg.

Today, the sun, the heat, the winds, drying out the damp forest, had lured the fire out of hiding. The ground fire crept up the dried underbellies of the spruce, licking at the old man's beard, that green, hanging lichen, and grew into a surface fire. Incited by the wind, the flames leapt up into a crown fire, producing a mammoth, dark, ugly smoke.

I picked up the phone and called Ralph.

"Holdovers can really come out of nowhere," he said. "Back in the nineties, a holdover blew up just east of your tower. The lookout went down for lunch. When she came back up, the fire was already at ten hectares. It blew up into a four-thousand-hectare complex."

Ralph once got caught off guard by a holdover that exploded after fifteen days of the fire burning low, hidden out of sight, along a stream bed.

"We can't catch them all," he said, sensing my defeat. "There's nothing you could've done differently, Trina. We're humans, not machines."

For the rest of the day, I watched the progression of the firefighters' efforts. The smoke disintegrated into a slight haze. By dusk they'd called BH on the fire, although it would take another week of extinguishing hot spots before a crew leader would announce EX.

At 8 p.m., I climbed down the ladder, feeling lethargic, frustrated. I cooked dinner and prepared my lunch for the following day, hoping the phone would break the silence. I was eager to chat with Jay, unload my frustration, and laugh off the day's events. I dialled his number. No answer. A soundlessness fell over the cabin.

With Holly curled at the foot of the bed, I climbed beneath the covers and read by candlelight until the words blurred, moving like ants on the page, and I fell into the clutch of dreams.

CHAPTER FIFTEEN

I rolled out a large oblong slab of grey clay to make a mug for Jay and wandered over to a stand of lodgepole pine to harvest the new shoot of green needles off the top of an immature tree—perfect for pressing into soft clay, for mimicking nature's patterns.

Working with clay was therapeutic and tangible. It relaxed my nerves. Making something physical, something practical, a drinking vessel, warded off my existential worries.

After sixty-some days alone, I had become, once again, vulnerable to those spiralling thoughts.

Who are you?

What are you doing with your life?

What about love?

What do you have to offer?

I remembered last summer, the breaking point after I ended my relationship with Akello, and the long gaze down from the tower. I didn't want to wander back into that dangerous territory. Somehow, working with clay grounded me; making mugs helped me to survive. In two months I'd fly back to Peace River, where I'd fire the mugs in an electric kiln. I hoped that Jay and I would drink from them together, maybe on a canoe trip along the Peace River, camping, cuddled up around a fire, beneath the stars.

The phone rang, breaking the morning in half. The mug was nearly finished. I leapt up from the picnic table and dashed inside, my hands slathered with wet clay.

"Hi," Jay said.

"Hey," I said, surprised to hear his voice. "What's up?" It was unusual for him to call so early.

There was a long pause. A silence—*real* silence.

"I'm calling because I need to tell you something," he said.

His voice sounded distant, far away, more so than usual. A knowing feeling crept up my spine. Instinctively, my defences went up and I braced myself.

"Is everything okay?"

"I slept with another woman," he said coldly.

Fuck.

My heart recoiled.

"And, well, I know we've been kind of talking about seeing one another in the fall, but . . ." His voice tapered off.

I considered his words. *Kind of talking.*

I wasn't sure what I felt more stunned by: that he had sex with another woman or that he'd just chalked up weeks of intimate conversation to nothing more than a casual friendship. *Kind of talking.* Well, I was *kind of* falling in love with him. I had sensed that he felt the same, but clearly I was wrong. Resentment flared up in me and I spit fire.

"Why are you calling to tell me *this*?" I asked sharply.

I heard the air suctioned out of his lungs.

"Has it not occurred to you where I am right now? That I'm alone in the middle of nowhere? Oh, and how about that you were the one who pursued *me* out here? What the fuck do you want?"

More silence.

"I'm sorry," he stammered. "I just felt that I should be honest with you."

"Yeah, well, it's all starting to feel pretty damn manipulative," I snapped. "I get it. I was like, *what*, a conquest for you? See if you can charm the Tower Girl? Just do me a favour and don't call anymore. No gifts. No letters. Honestly, just leave me alone. I want to be alone."

I hung up before he could say another word.

"Fuuuuuuuuucccckkk!"

My lungs emptied, I threw myself outside onto the sloping lawn, tearing out handfuls of grass, sobbing face down in the dirt.

"I'm such a stupid woman," I wept.

I could see myself there: how pathetic I must have looked, how wildly out of control. But what did it matter? There wasn't another soul around to see.

If a tree falls in the woods and no one is around to hear it, does it make a sound?

It was the kind of philosophical question that, to me, revealed the height of human arrogance. Were we so proud a species as to think that our presence mattered that much in the wild? That only *we* could give legitimacy to the forest's song? Does wildfire not crackle through old pine? Does a caribou not bray in pain when pulled to the ground by a wolf? I knew this much to be true: if a woman lost her mind at a fire tower and no other human was around to hear, she made a sound. A very loud, very animal, very terrifying sound that went like:

"AWWWWWWHHHHHHHHAAAAAAAAAGGGGGGGGGGGGGGGGGAAAAAAAAAAAAAAAAA!"

I needed to get out. I grabbed a backpack, some water, and a granola bar, and threw myself blindly into the forest. Holly bounded ahead, excitedly sensing a journey, but there was nowhere to go. Muskeg to the south, north, west, and east. I wanted to run away from the tower, from being a lonely, foolish Rapunzel archetype. I wanted to be strong. I wanted to be a woman who didn't need rescuing.

I stomped down a trail leading north, plodding through long, waist-high grass that was slick with droplets of moisture. Within seconds I was soaked to the bone, my jeans heavy, wet, stuck to my skin. Brown water sloshed around in my rubber boots. My eyes blurred with hot tears.

I couldn't even see the trail ahead. I knew that I was breaking all of the hiking rules:

Always let your tower buddy know where you're going. *Fail.*
Take a first aid kit with you. *Fail.*
Be alert to the trail. Look ahead. Listen. *Fail. Fail. Fail.*

The trail was a sea of green. I could have tromped right past a bear or moose, completely unaware. Then again, what creature would dare to mess with Tower Girl, screaming bloody murder? I imagined the wild-life in comic form, cowering in the bushes, their wide eyes watching me crash by. A speech bubble emerging above a startled black bear's head—WTF??!!!!!

Who was she? I festered. What did she look like? How had they met? Jealousy seized in my chest as I conjured up the image of Jay and the woman together, their limbs passionately intertwined. Was she pretty?

Stupid, stupid woman. I thought I'd been looking out, finally seeing the world clearly, but Jay had been another optical illusion, a trick of light. I'd fallen in love with a mirage. I'd constructed a fantasy from a bouquet of pussy willows, a letter, a voice, and the clouds outside my cupola window.

And yet . . . he had called to tell me the truth. I could hear his shame on the other end of the phone. My violent sobs softened and I began to sort more rationally through my thoughts. With whom was I really angry? Jay, or myself? Who was really the villain? At least he'd been honest, whereas I'd kept my infidelity a secret from Akello for 365 days.

Stupid, stupid woman.

WHO-WHO-WHO!

My neck snapped up. Suddenly, a massive bird, with wings spreading four feet wide, swooped low, so low I heard the air swooshing through its fanned feathers above my head.

Oh!

The bird perched ahead on a tall pine. Its head spun nearly 360 degrees and it glared down at me with yellow eyes. A great horned owl. I sucked in a startled breath and suddenly forgot everything.

Our eyes locked in a stalemate. Who would look away first? But the owl had seen enough. It leapt, talons curled around nothing, wings spread onto the wind.

SWOOSH, SWOOSH, SWOOSH!

"Come back!" I cried to the owl.

I looked down at Holly. Her fur was slick with moisture. Mosquitoes buzzed around her face. Her tongue was out, a dangling smile, and her black-and-white chest heaved up and down. She was indifferent to my anger and grief. Joyful to be out exploring, threading her body through the wild.

I glanced down the trail. We could keep going north, another twenty-five kilometres, until we hit a logging road. I could throw in the towel on the season. Check out. Instead, I checked the hour on my cellphone: almost noon. The radio dispatcher would soon be waiting to hear my afternoon weather report. If I didn't answer her call, they'd be anxious. The absence of a voice—*silence*—would alarm everyone around me. My neighbours' ears would perk up and everyone would worry: What happened? Within minutes the duty officer would send a helicopter and firefighting crew to investigate.

Decades ago, a lookout had walked away from her cabin in the middle of the night, informing no one, and travelled deep into the bush. A helicopter found her the following day, lost, alone. She was disoriented, but walking. Where was she going? I'd heard the story at the training in Hinton before flying in for my first fire season. Crazy, I'd said at the time. That's totally crazy. Why on earth would she do that? Take herself off the map like that?

But what do we really know about the nature of isolation? What do we know about survival?

What is the longest you've ever been alone?

Holly looked up at me with those large amber eyes of hers. I sensed that I was dangerously close to an edge, a limit. But also, I knew that I couldn't just turn my back on the fire season and walk away. Real lookouts—the ones who become lifers—had learned how to endure emotional hardship alone. Broken hearts, mistakes, regrets. Shame. No matter the obstacle, skilled lookouts didn't just walk into the bush. They got back on the ladder. They climbed.

Holly wagged her bushy tail.

I followed the black-and-white dog back to the yellow cabin. Just in time for the twelve o'clock ladder check and p.m. weather report. I swung my faded red harness over my shoulders, looked up towards the cupola, and pulled myself up on the ladder. Skyward, I climbed.

The phone stopped ringing, and I was left, once again, alone, with the nature of my own thoughts. How badly I wished to write my grandfather a letter and ask for advice on matters of the heart.

I thought of the summer he taught me how to carve. I was twenty-four years old.

In 2009, I was leading a humanitarian construction project in northern Nicaragua when I learned that my nana had died, swiftly, from a twisted intestine. I missed the funeral, but travelled back to Canada a month later and flew to Regina to spend two weeks with my grandfather. He promised to teach me how to carve.

I knew I wasn't really there to learn how to carve wood. The grain moved much deeper than that, though my twenty-four-year-old self wasn't quite ready to grasp the meaning behind his stories and wisdom.

Every morning, we woke and drank coffee in our pyjamas and shared stories at the kitchen table. My grandfather was grieving the loss of my nana. They'd been planning to celebrate their sixtieth wedding anniversary and had even mailed out invitations for the party, but she died shortly before it was to have taken place. He moved about the small duplex they'd shared together, feeling her ghost in every room. He showed me the black-and-white photograph of her at only sixteen or seventeen years old that he carried with him through WWII. Once wedged in front of his air gunner seat, the photo now hung from the wall in his woodworking shop.

"Your nana was the love of my life," he said.

He bought me a carving knife, an electric drill and sander, and a wood burning tool. We worked outside in the backyard, enjoying the sun on our skin as he showed me how to carve a bird's feather from soft wood.

"You have to listen to the heart of the wood," he said. "Feel for the grain and work with it, not against it. You have to discover what it wants to become."

We drank copious amounts of coffee together every morning. I got up to make the first steaming cup of the day and hand-delivered it to my grandfather, propped up in bed. He turned on the morning news, though we didn't really watch or pay attention to what was happening in the world beyond. I sat down on the carpeted bedroom floor and listened to my grandfather remember, aloud, how the story went.

I felt like a small girl again, hanging on his words, only I hadn't heard these stories before. He told me about a near-death experience while flying over the prairies in Manitoba. After WWII he got his pilot's licence, wanting to soar amongst the clouds, and found a job flying a small aircraft, delivering farm equipment on the Prairies. One day, one of his engines failed and the plane started to nosedive towards the ocean of wheat below. He grabbed a parachute from the back of the plane and the photograph he kept of my nana, the same one he'd carried in the war.

"I remember seeing her face," he told me. "I didn't want to lose everything we had together. I had so much to lose. I began to say the Lord's Prayer aloud. Then I jumped."

He told me about a friend named Ben, one of his platoon mates during WWII, who'd died only days before the war came to an official end. It had been a rainy day and the young man had slipped and split his skull wide open. They dug his grave with the rain falling hard and everyone put on an article of Ben's clothing, as was the tradition, to honour the loss of their friend.

And then he told me something that I didn't quite understand, but it would resonate deeply with me now, after he was gone.

"When I look back on my life, I can see how it's easy to make mistakes. Making a mistake happens in the blink of an eye," he said slowly. "But letting go of the shame of hurting someone you love—that can take a lifetime."

I wished I could go back now and ask him about the source of his shame, somehow release him from it. Whom had he hurt? And why had his pain stayed with him for so long?

"Tell me about your shame, Grandpa," I'd ask him. "And I'll tell you mine."

PART FIVE

RE-GEN(ERATION)

I could spend a whole day listening. And a whole night. And in the morning, without my hearing it, there might be a mushroom that was not there the night before, creamy white, pushed up from the pine needle duff, out of darkness to light . . .

—ROBIN WALL KIMMERER, *Braiding Sweetgrass: Indigenous Wisdom, Scientific Knowledge, and the Teachings of Plants*

CHAPTER SIXTEEN

Now what? I demanded from the forest.

Now, nothing, responded the wind through the leaves.

A spider scuttled across my arm with her silken legs. But I don't want to be alone, I told her.

She said something, but I couldn't yet speak her language.

Tell me about your loneliness and I'll tell you mine, I said to Holly. She licked my hand.

What is the cure for loneliness? I asked the moon.
    Solitude, she whispered back.

In 1993, **Maurizio Montalbini**, an Italian sociologist and caving explorer, descended alone into a cave near Pesaro, Italy, for 365 days as part of NASA research on the impact of isolation on human health. Montalbini survived on high-calorie pills and astronaut foods and broke his own world record for enduring complete isolation. He read nearly one hundred books. "One cannot fight solitude," he said. "One must make a friend of it." When he came back up to earth, he believed he'd spent 219 days underground, when in reality it had been one full year. "I carried everything inside me—my love, conviction, ideals," Montalbini told reporters after he emerged from the dark. What did he miss the most while underground? His family and friends and the "taste of cheese."

Edith Bone, a Hungarian journalist, was a political prisoner in solitary confinement in Budapest from 1949 to 1956. She survived the isolation by making a mental inventory of all the words she knew of the seven languages she spoke. She wrote poetry using bread crumbs from the hard, stale rye bread she was given. She also used the crumbs as an abacus, counting the long days of her incarceration. In her autobiography, *Seven Years Solitary*, Bone wrote about her exhaustive efforts to see beyond her prison cell door. Bone removed single threads from towels and wove them into a rope, which she used to wiggle loose a large nail in the heavy oak door of her cell. She sharpened the nail on the floor, then drilled a tiny hole through her cell door so she could peer through a pinprick hole of light.

Steven Callahan, a sailor, was forced to abandon his vessel, the *Napoleon Solo*, while journeying to Antigua in 1982. As his boat was overwhelmed by the breaking seas, Callahan escaped into an inflatable life raft with only a few survival tools: a speargun, flashlight, flare, and sleeping bag. He survived seventy-six days alone, adrift at sea. "My life went by my eyes very slowly," said Callahan. "I regretted all my mistakes and errors." He kept a navigation log. He connected with the ocean on a spiritual level. "I became very attached to the fish," said Callahan. "They were kind of symbolic of the magic and mystery of life and the sea." The fish fed him, and nearly killed him. He survived several shark attacks. "And in the final analysis, they brought my salvation." The fish drew in the birds, hovering over his raft, which attracted the attention of nearby fishermen.

Lucas Miller, thirty-two years old, signed up for a reality TV show called *Alone* in 2015. Miller and nine other contestants were dropped off at remote locations in northern Vancouver Island, armed with ten survival tools. The goal? Survive as long as possible, alone, in the wilderness, while filming yourself with a hand-held camera. Miller spent a month constructing a yurt, fireplace, canoe, and paddle. He fished, hunted, and foraged for food. After thirty-five days alone he carved a rustic guitar out of wood, burning out a sound box with coals and stringing the instrument with fishing line. "The isolation, it's unlike anything I've ever experienced," said Miller, zipped up in his sleeping bag. "I can't run away. I have to look at my life." He wiped his tears with a faded red bandana. "It's about accepting who I am." On day thirty-nine alone, Miller tapped out of the contest. "If you keep worrying about how good you are in comparison to other people, you'll never win."

Trina Moyles spent four consecutive months alone in northwestern Alberta working as a fire tower lookout. "Don't you get bored out there? Don't you get lonely?" people often asked her. "Yes," she said. "And yes, it's the most beautiful kind of loneliness that I've ever known." Moyles was saved by the act of observation. She watched the incremental change of light over the forest. She watched clouds and birds and beetles and the ripening of blueberries. Nature was a nectar. When the forces of her life beyond the tower felt too overwhelming to consider, she meditated on the wild minutiae around her: a hummingbird moth on a dandelion, the piercing cry of a kestrel, the fireweed erupting with pink flowers. She wandered the forest with her dog, Holly, who became more than a dog—a friend, a colleague, a love. Holly taught Moyles about how to stay grounded in the moment, how to read the wild and let go of the past. Especially, how to play and laugh and stop taking life so seriously. "What do you do out there?" was another question that people enjoyed asking her. She would tell them: "I read books, hike, sew quilts, do pottery, and write." But the truth was that she survived by opening her eyes.

Writing saved me too.

I wrote up in the cupola, typing out a few sentences then looking up to scan the forest every sixty seconds or so.

"What a great place to write," people often said to me, not realizing how easy it would be to miss a smoke. They didn't know that the winds could blow at fifty, sixty kilometres an hour and how the cupola could shake back and forth like a rattle. When wildfires blew up, the radios sizzled with energy and there was rarely a quiet moment. People also didn't appreciate the terror of writing in a one-hundred-foot steel structure during storms—the extreme probability of being struck by lightning. I wouldn't tell them about the no-see-ums that swarmed up in the cupola during late May, nor the bird-sized mosquitoes that preyed upon me when I climbed down the ladder every night. I definitely wouldn't tell them about pooping in a bucket or flinging my urine out the window. How romantic, indeed.

There were better places to write, but I wrote anyway.

In the northern boreal, during the peak summer months, sunlight lasts until midnight. In the evening, when the risk of anything in the forest burning was low to non-existent, I climbed back up the tower with my laptop stowed in my backpack. At 9 p.m., I sat up in the sky, watching the long northern light start to dim and paint the trees into shades of green gold. At Hinton a ranger had told me, "Climb when the sun is low in the sky. You'll be able to see the trees, the lakes, the hills and roads—you'll see everything—with greater clarity and detail. It's the best way to get to know your forest."

From the sky, I wrote with abandon, emotionally stirred by the shadowy golden light, the illuminated forest, and the growing sense of connection I felt with being a lookout and a woman who'd come back to the North, to the landscape that had shaped me.

What I wrote, I wrote for me.

Writing was a way to converse with the past, the long forgotten, and even the dead.

At what point did my isolation soften into solitude?

I can't remember the exact moment.

It was like learning a new language:

For many days, your brain struggles, reaching around in the dark for nouns, verbs, tenses, and meaning. It's exhausting simply to understand and be understood. So you compensate by mixing languages and gesturing with your hands. Often you say the wrong words and confuse, or amuse, or offend whomever you're trying to communicate with. Awkward pauses and crossed wires and, sometimes, surprising connections occur. But at a certain point, you realize your brain is no longer searching for the right word. You understand. You feel the language beneath your skin, it moves your heart, excites your senses, and echoes loudly in your dreams.

Isolation is different from solitude. It's about being cut off from the world, or cutting oneself off. Isolation is akin to hiding alone in a dark room, or being trapped in a sensory-deprived state. Isolation can be forced upon us, or it's a state we impose upon ourselves. Last year, when my relationship with Akello ended, I felt as though I didn't deserve to belong in the world. I deserved to be lonely. Loneliness was a kind of punishment, a self-abandonment.

Solitude is about actively engaging with space; it's a voluntary commitment to self. It's about listening to your thoughts and trying to see with greater clarity. Solitude is a deepening connection with nature, finding belonging within the four walls of your own skin.

I can't tell you when or how it happened that second summer.

One day, I woke up and saw my solitude staring back at me.

Like a flower.

Like a bird.

Like a forest on fire.

In the early hours of July's full moon, my niece was born.

I woke to a text from my brother. A photo of him dressed in a yellow hospital gown, his weary eyes brimming with tears, holding my swaddled niece. Her moon face glistened. My brother, my protector, my best friend in the adventure of childhood, now a father of a daughter. I clutched my phone, gazing incredulous at the image of my brother and niece.

Brielle, they called her. Her skin was the shade of a wild rose petal.

I couldn't wait to know her.

CHAPTER SEVENTEEN

I leaned back in the padded swivel chair, my legs and bare feet dangling over the edge of the cupola window. Pods of clouds paraded across a placid blue sky. I strummed a few chords on my ukulele. I was trying to learn Leonard Cohen's "Hallelujah" chord by chord. The strings twanged and echoed off the cupola walls. Holly skulked beneath the solar panel. She loathed the ukulele. When I'd play up in the cupola or down on the ground, she'd hang her head and slouch away. "Sorry, girl," I murmured.

The phone rang. Was it Jay? I swatted away the thought like a horsefly. Stop thinking about him, I scolded myself. Ten days had passed since that morning he'd called. Even though I was feeling the rhythm of my solitude at the tower, I missed our long conversations, debriefing my days with him. He'd made me laugh.

But it wasn't Jay calling. It was Justin, a fellow lookout from a tower located over a hundred kilometres to the south. We spoke on the phone every few weeks, updating one another on lookout news related to weather, wildfire, and wildlife. Justin was related to Holly's former owners.

"How's she doing?" he asked.

"Ah, she's a happy girl," I said glowingly. "The love of my life out here."

"Glad it worked out for you to take her. That was a tough call to make."

"What do you mean?"

"They didn't tell you?" he asked, then hesitated for a second. "They were going to put her down."

"No, wait, *what?*"

"Holly killed their dog," Justin said, shocked that I didn't already know. "Like, we're talking that she killed the family pet. One day the two of them got into a scrap. Holly hamstrung her—you know, the way wolves pull down a moose. She went for the dog's hind legs. When they came home, they found Holly on the dog's throat."

"*What?*" I said, incredulous. "I didn't know anything about this."

"Yeah," Justin said slowly. "The family was really torn up about it. Half of them wanted to put Holly down. The others thought they should find her another home. And then, well, you came along."

I couldn't fathom it. Holly, a murderous beast? I glanced below. She was out from under the solar panel and lying in her favourite spot beneath the fire tower. I traced with my eyes the white spot on her belly.

"Are we honestly talking about the same dog?"

"She's practically a domesticated wolf."

Justin told me he used to go hunting with Holly on the farm, and of the three dogs, she'd been the most attuned to the wild. Hunting was in her blood. She was born on one of the First Nations reserves around Peace River, semi-wild, running with a pack of rez dogs that belonged to nobody and yet everybody in the community. People would shoot a deer to feed the stray dogs. When Justin's relatives adopted Holly, those feral instincts followed her to the farm.

"One day, the cops showed up at the farm," said Justin. "The neighbours had called to complain because two of their dogs had taken down a deer in the snow. The cops thought it was the big black dog, but we knew it was Holly. We saw the blood on her snout."

Those two words: *domesticated wolf.*

I tried to imagine Holly, her chest heaving from the exertion of pulling down a deer, feasting on the open belly, blood dripping from her mouth onto the white snow.

I thought I knew everything about her nature. Joyful. Loving. Loyal to the bone.

When I climbed down the tower, Holly was waiting for me, her white-tipped tail wagging like a flag. *Arrooooooooooo!* she called happily. I looked down at her and wondered, Would I have adopted her had I known the truth about her past? Probably not, even though I'd wanted a dog that would protect me. A fighter. But few people would adopt a dog with a story like hers. I was relieved that the owners hadn't told me.

I couldn't imagine a day without her. Holly was more than a dog, she was a comrade.

A saint and a wolf.

Akello's face materialized on the computer screen: a wide, happy grin. He laughed loudly. My heart melted at the sound of his laughter. It was so good, but so strange, to see him again. Many months had passed since we'd come face to face. Nearly two years had passed since we'd physically been together.

"Hey Treen!" he sung to me.

"Hey Ake," I said softly, a hard lump forming in my throat.

Akello's life was full of familiar activities and new ones too. He was working at the automotive garage every day and playing football with a men's league in Kabale. He used the money he'd saved to invest in a boda boda (motorcycle taxi) business. He was working for a local NGO to build rainwater catchment and storage systems in several nearby villages. Months ago he'd moved out of the beautiful apartment we'd shared together.

"What do you want me to do with your things?" he asked.

"I don't need anything, Ake," I told him. Then I remembered. "I'd like to keep the dress, though."

The shimmering turquoise cloth that I wore the day we didn't get married.

"I'm sorry." My voice broke suddenly. "I am sorry I ended things the way I did last summer. I was completely stressed. I couldn't take the pressure anymore. I'm sorry I hurt you."

He chose his words slowly, a proverb formulating on his tongue.

"Oh, Treen," he said tenderly. "I think it's good to forgive because no man is perfect."

"But I am so weak," I sobbed. "I am a bad woman. I fucked everything up."

"No, you are a very strong woman," he said. "We were happy together here in Uganda, but reality got in the way. I think my heart was designed to love you, because I've never stopped loving you."

The tears came hard and fast. Holly looked up from where she slept on the floor.

"Trina," Akello said softly. "Don't you know that everything happens for a reason?"

Me: What makes a good lookout?

Ranger: A good lookout is observant, vigilant, self-disciplined, resilient, and—

Me: Okay. But what *really* makes a good lookout?

Ranger: Over the years, I've met lookouts from different walks of life. Lookouts who are doctors, priests, former sex-workers, bartenders, nuclear physicists. Do you know what they have in common?

Me: No. What?

Ranger: They've learned to be all right with themselves. If you want to survive out there, then you must be enough for you.

———

Notes to self:

Nature suffers, as the body suffers.

Nature is not indifferent. Nature needs love to endure.

So do you.

And given that you are the only human in a forty-kilometre radius, you must learn how to show compassion to yourself.

Your survival depends on it.

No one else can save you out here.

Surrender to your nature.

Learn how to be tough enough—tender enough—to endure another day.

Record high temperatures in June rapidly sucked the moisture from forests and grasslands in the southern and eastern parts of British Columbia. On July 6, lightning struck and the earth was primed to burn. Over one hundred fires ignited within a few hours, candling and torching entire trees. The small fires bled into larger fires, and the larger fires created raging complex, or campaign wildfires. On July 8, 2017, the Government of British Columbia declared a provincial state of emergency, calling for national and international assistance with hundreds of wildfires burning out of control.

Five thousand people worked on the front lines of the wildfires, based at makeshift camps throughout B.C. to battle the blazes. Alberta answered B.C.'s plea for help, exporting hundreds of firefighters, wildfire rangers, radio dispatchers, and personnel. The Canadian Armed Forces contributed aid, and firefighters from Australia, New Zealand, Mexico, and the U.S. joined the efforts.

Twenty separate fires were burning on the Chilcotin Plateau in the Cariboo region of the province. No one knew it yet, but the fires would amalgamate to form the Plateau Complex. The Plateau would become the largest fire in B.C.'s history, burning nearly 550,000 hectares of land, nearly as big as the Horse River Fire. Nearly as big as Prince Edward Island.

"I was so afraid that I'd never hear from you again," said Jay. "I know you're vulnerable out there and I'm sorry that I hurt you. But I needed to be honest with you. Even though we weren't really together, I-I felt like I . . ."

"Like you cheated on me," I finished for him.

"Yeah." He paused. "Trina, I've never talked with anyone like the way I talk with you."

I stayed silent, though I felt the same, and it was the reason I'd called his number. I swore to myself that I'd cut him out of my life, but I couldn't. Akello had forgiven me—could I forgive Jay? I missed him and the way we used to talk until the midnight sun went down. I missed the voice I'd fallen in love with.

"Are you with that woman?"

"No," he said quickly. "That was just a drunken hookup. It honestly didn't mean anything."

Those words again, *it didn't mean anything*. I'd spoken them before. A red flag, a warning. Should I trust him? Did I even trust myself? My guard stayed up. I was in a tower, for fuck's sake. I wouldn't fly home for another six weeks.

"I want to see you," he said. "If you'll let me."

"What? How?" I asked, skeptical. "You mean in the fall?"

"I was thinking," he said slowly. "I could hike in next week on my days off. If you'd let me."

How does the saying go?

No man is an island.

No woman is a fire tower.

CHAPTER EIGHTEEN

I glanced down at the clock on my phone: 10:13 p.m. The sun was dropping low over the western horizon—only an hour and a half left of sunlight now. I stared hard at the edge of the bush, as though Jay would emerge any second. It was improbable for me to imagine that he was somewhere out there following a game trail, bushwhacking, and sinking waist-deep in muskeg, following a precarious compass bearing up and down steep valleys to reach me. He'd set out nearly seven hours ago for the thirty-kilometre hike.

I grabbed my hand-held radio, slipped my harness over my shoulders, and climbed up the tower. The low, golden light illuminated every needle on the spruce and pine, every leaf of the aspen and birch. I looked north at the dense cover of bush. The boreal seemed to go on forever.

"JAAAAAAAAAYYYYYY!" I screamed at the top of my lungs. I held my breath, waiting for a response. A songbird trilled. Holly craned her head skyward, wondering what the fuss was about.

Worst-case scenarios looped violently through my imagination.

He'd tripped and badly sprained his ankle and couldn't walk.

He'd lost the compass bearing and was walking off course.

He'd been stalked by a cougar.

He'd crossed paths with a sow grizzly and her cubs.

His eyes had been punctured by a branch and he was walking around blindly in the woods.

I turned on the hand-held radio and tuned in to Channel 20, the frequency we'd agreed upon. I pressed the transmitter button once, twice. Silence.

"Hello?" I heard Jay's voice crackle through the radio.

"Oh my god, I'm glad to hear your voice! Are you okay? Where are you?"

Silence.

"I just passed an old well site," he finally said. "I'm not sure how much further I have to go. Maybe five or six kilometres? Can you hike in to meet me? I'm exhausted and losing steam."

"Okay," I nervously agreed. "But keep your radio on."

I packed up a bag of water, snacks, a flashlight, and a radio, then put the canister of bear spray on my belt and slung the shotgun around my shoulder. I stuffed a sweater into my bag and glanced at the phone. I should inform someone, I thought. It was risky to hike around at dusk. Many predators were more active when the sun went down and the heat of the day cooled off. And what if I got lost? Someone needed to know where I was going. No one even knew Jay was hiking in to the tower.

I texted a friend in Peace River.

> Jay is hiking out to see me. I'm going to meet him on the trail. If I'm not back by 1 AM, can you let someone know?
>
> He's hiking in?! Isn't that like 30 km? Is he crazy?!
>
> Yeah. He's only 5 km away.
>
> Are you sure you should be hiking out right now? It's getting dark.
>
> I've got the gun and bear spray. And Holly is with me, too.
>
> Okay, but be careful. Text me when you get back.

———

Holly happily loped ahead as we moved north through the bush.

"Hey bear! Hey bear!" I hollered before every bend in the path. I held my breath, worried I'd come face to face with a grizzly or black bear. What would Holly do? Charge the bear? Whatever happened, she would defend me, I decided. Now I knew what she was capable of.

The grass grew so tall and lush that occasionally I lost sight of Holly. Her white-tipped tail emerged every so often as she threaded her way through the fuchsia fireweed. At one point I looked down and noticed a perfect white-and-brown-checkered feather on the ground in front of me. An owl feather. I picked it up and ran the soft feather tip along my lips. A good omen. I thought of my grandfather John and the feathers we'd carved together out of wood.

The sky was fading fast, turning a faint blue-lilac shade. I glanced down at my watch. We'd been hiking for nearly forty-five minutes. No sign of Jay.

I clicked the radio transmitter.

"Hey, can you hear me?" I said into the radio.

"Yeah, I've got to be getting close now." His voice came through louder, clearer.

He came out suddenly from beyond a curve, three hundred metres ahead on the path, wading through waist-high grass. It was startling, bordering on disorienting, to see another human, someone who had not flown in.

He looked completely exhausted, bent forward, eyes down on the trail, oppressed by the weight of the monstrous pack strapped to his back, a rifle slung around his shoulder. He wore a bright-blue long-sleeved shirt and a sand-coloured safari hat. He hadn't yet seen us.

Holly stopped dead in her tracks. She couldn't see above the grass, but she sensed something approaching. She sprang a metre off the ground with all four legs, frantically trying to see beyond the grass, as if she couldn't fathom the possibility of meeting another human out here either. Humans came from helicopters. They did not materialize out of the bush.

Jay looked up suddenly. He flashed us an exhausted but relieved grin. Holly leapt ahead.

My whole body trembled. I had once said that if a man appeared out of the woods, suddenly, unexpectedly, I'd make sure he was staring down the barrel of my shotgun. But I felt a shot of love when Jay's familiar form emerged from the wilderness. My defences peeled off like snakeskin.

"Hi," he said, sweaty, exhausted.

"Hi," I said, suddenly painfully shy.

His shirt was soaked through with perspiration. His knees buckled under the weight of the bag. It rolled off his shoulders and thumped the ground.

"Come here," he said tenderly.

We embraced. It wasn't electric, or fire. It was water. It was relief. After one hundred days of solitude—save for the days when the firefighting crews lounged beneath my tower—and relying on myself for every little thing, it was the purest kind of relief I'd ever felt.

"You crazy, crazy man," I said. I pulled the owl feather from my pack and tucked it into the mesh siding of his safari hat. Then I grabbed his heavy rifle. Two guns on one shoulder.

"Are we close?" he asked, a pained expression on his face.

"Almost home," I said.

I took Jay's hand and led him back to the yellow cabin in the woods.

He was so loud. He sang and told bad jokes, and his deep voice shattered the silence of the cabin. On the one hand, I was elated to have another human being within arm's reach. On the other, *he* was a strange presence in *my* space. Even Holly seemed out of sorts. She kept looking at him curiously, sniffing the air, trying to decide who and what he was and why he had come.

"Are you sure you don't want me to boil you water for a shower?" I asked him.

He was peeling off his sweaty hiking clothes behind the bedroom door. I heard him filling up the shower bag to rinse off. I caught a glimpse of his bare, muscled back and looked away.

"AHHHHHHHHHHH!" he shouted, the cold rainwater shocking his hot skin.

I was frying eggs and bacon in a pan and slicing thick pieces of my sourdough bread when he came out of the bedroom, his long hair slicked back, dripping wet. Shit, I thought, he's so handsome.

"Well, now I'm awake," Jay said jokingly. He sat down at my kitchen table and scarfed down the eggs and bacon and buttered toast as Holly and I both gaped at him. He asked about my fire hazard level and schedule for the following day. The duty officer had designated my tower on high hazard and I'd be up in the sky from 11 a.m. to 8 p.m.

"Do whatever it is that you need to do here," he said, inhaling a piece of bacon. "I don't want to get in your way. Just pretend I'm not here."

"Umm," I laughed. "I don't think that's possible."

Once he'd eaten and we'd cleaned up, he pulled a sleeping bag from his pack and unfurled it on the cabin floor.

"What are you doing?"

"I didn't want to make any assumptions," he said. "I want you to be comfortable."

"You did not just hike through the bush to sleep on my living room floor," I smirked at him. "You can sleep in the bed." Please, I wanted to beg him. For god's sake, sleep in my bed.

"Are you sure?" he asked.

"Yes, I am sure."

Jay crawled beneath my quilt on the mattress made for one and my whole body quivered, thinking, Oh my god, there's a fucking man in my bed. I lit a candle and turned off the generator. The electricity fell away, leaving us in the soft, golden glow of a single flame. I climbed under the covers and he turned his long body to face mine. We smiled at one another like two giddy kids with a secret. And then I reached across that last bit of distance that remained between us—just a few centimetres of unknown wilderness—and kissed him.

Jay slept most of the following day, his legs stiff and aching from the long trek through the bush. I kept stealing down the ladder throughout the day, peeking into the bedroom, floored to see him curled up in my bed. I felt as though I was conjuring it all up, a wild hallucination, an epic trip down the rabbit hole. I watched his chest rise and fall, Holly curled up next to him, uninterested in her duties as Tower Dog. Who could blame her? Suddenly, watching for smokes felt incredibly tedious. I hoped that nothing would burn. All day I floated on a cloud. After lunch Jay emerged from the cabin, shoeless, shirtless.

"Good morning!" I hollered down at him. "Do you want to climb up?"

"I don't know. My legs are pretty sore, and . . ." He looked up doubt-fully. Was that a look of fear? "I thought I could do some cleaning and yardwork for you."

"That would be amazing!" I yelled down. "Tools are in the engine shed."

He rolled a cigarette and sat down at the picnic table and smoked. Then he got up, clambered around in the shed, and began weed whack-ing around the perimeter of the yard. He even got down on his hands and knees and dug up the stubborn weeds growing between the cracks in the concrete sidewalk blocks that led from the cabin to the foot of the tower. Later that evening, when I climbed down and entered the cabin, Jay was at the stove, pushing around pork chops in the frying pan with a spatula. He was playing a catchy pop song on his phone.

"Welcome home," he said.

"Oh my god," I said, ravenous and relieved. "You can stay forever."

"Oh yeah?" he teased. "Maybe I could be your Tower Husband."

I hadn't made or shared food with another man for so long now. Akello's words rang in my ears: "It isn't good to eat alone," he often said.

Jay and I fell into a rhythm of cooking dinner together while Holly watched hungrily. We sat down at the round wooden table to eat.

"I can see why you do it," he said to me.

"What?"

"This job," he said, rubbing his beard thoughtfully. "It's peaceful out here. I haven't been able to hear my own thoughts for the whole fire season. Camp life is great, but it's tough. There's nowhere to go, really. It's hard to find the space to think."

"Yeah, I know," I said. "But some days there's too much space to think. I've had some really hard days out here. Last season . . ." I didn't finish the sentence.

"Trina, I really am sorry for what happened," said Jay, putting down his fork. "I wanted to hike out immediately after you hung up on me. I didn't think you'd ever want to talk to me again."

"I don't think that would've been a good idea," I said coolly. "I was pretty upset about everything. You don't know what it's like to be out here. To get that kind of phone call from someone you care about."

"I really care about you," he said softly. "That was a stupid mistake."

"It's okay," I said. "I'm just happy you came, though you're totally nuts."

"I hope I'll be able to walk normally again," he said, laughing, revealing that slight gap between his front teeth that I adored.

My hand found Jay's hand. Such a simple act: holding the hand of someone you care about. And how good it felt to be held by this man who'd hiked thirty-some kilometres through swamp and bush to share space with me. How good it felt to be vulnerable again. I told myself that I wouldn't ever take it for granted.

Jay was waiting for a call back to work and only planned on staying for a few days. The wildfires that had ignited in B.C. last week—the beginning of the Plateau Complex—were growing increasingly out of control. Authorities were evacuating many communities, including nearly twenty-five thousand people in the Williams Lake area. Jay and his twenty-man unit crew were anticipating an export into central B.C. to help contain the flames.

He kept checking his messages, standing up on the picnic table, holding his cell up to the sky, searching for a couple of bars of service. I kept bracing myself for his departure. Don't get too attached, I warned myself. You have another six weeks of isolation ahead of you yet. But the call back from his crew leader never came.

Jay stayed with us for nearly a week. We fell into an easy rhythm together. I kept on top of my lookout duties and he made life considerably easier, helping with meals, hauling rainwater, washing dishes, and sweeping the cabin. He rearranged my pots and pans cupboard and cleaned out the disastrously messy engine shed. One day I managed to convince him to put on the other harness and climb up into the cupola with me. He'd been trained by Forestry to climb towers, but even so, it was evident he disliked heights.

"It always feels way higher than it looks!" he screamed, frozen on the ladder about halfway up.

"You're doing great! Use your legs! Just look straight ahead at the rungs!" I called down to him.

He burst through the cupola hatch and it occurred to me that I'd never had another person up in the sky with me. The cupola was my sacred territory, a nest built for one.

"Wow," he said, breathless, staring out into the wide open space.

We played a few hands of crib, devoured turkey and cheese sandwiches, and watched large cumulus formations drift by. A storm was slowly building to the far southwest.

"I don't know how you do this," he said. "All day, every day. I would go crazy up here."

"You just . . . I don't know how to explain it . . . you adapt," I said. "You sink into it."

Dangerous-looking thunderheads approached from the south and spread out like a purple curtain across the southern horizon. Predictably, the storm cell split apart. One cell broke off, moving east, the other west, both cells tracking directly around the tower. Lightning was imminent. I instructed Jay to watch the western cell while I kept my eyes glued to the east.

"First strike!" we yelled at the same time.

"Do you want to call it in?" I asked jokingly.

We watched for another several hours, waiting for a smoke to pop up where lightning had struck the forest, but nothing happened. We shared the one chair in the cupola. I sat on Jay's lap, leaning back against his chest as he wrapped his arms around me. The sky dropped and coalesced into a heavy woollen blanket that we pulled over our heads. Rain pattered gently against the windows.

Loving him was so easy.

Jay and I patiently picked enough tiny wild strawberries from the sloping front yard to fill a measuring cup and made pancakes with the berries sprinkled on top. He helped me wash the laundry by hand and lay the clothes out in the sun to dry. We played cards. I taught him how to make a loaf of sourdough bread. He taught me how to fell a black spruce tree. We sawed the tree into logs and made a garden bed beside the yellow cabin and planted my favourite herbs. We read books and strolled through the meadow of wildflowers. Holly was smitten. She began to follow Jay around like a shadow, as she formerly followed me, always eager for a pat, or a run, or a bit of rough play. It was obvious that he adored her too. We were a pack of three.

Jay didn't pry into my past. Our love that week was less about words—we'd spent weeks and weeks talking over the phone—and more about actions: Mixing together the ingredients for bread. Pushing seeds beneath the soil. Reaching for one another.

Saying goodbye was unbearably hard. I was already nostalgic, because I worried that when I saw him next, after the fire season had wrapped up, after I'd flown back to Peace River, our relationship might not feel the same. Life at the tower was, indeed, a kind of fantasy world, a world sheltered from so many distractions: economic necessity, people, travel, opportunity. It was a bizarre but absolutely pure place to connect, heart to heart, with another human being.

I made him a lunch for the long walk back to civilization, and then prepared myself for his departure and my return to solitude. We hugged goodbye at the mouth of the trail that led north.

"Thank you for such an incredible week," I said.

"I love you," he whispered in my ear.

"I love you too."

We let one another go. Holly and I walked back to the cabin and I felt the physical ache of being alone again. How suddenly he'd burst into my territory then vanished into thin air. Had he even come? Maybe I'd just imagined everything.

His scent lingered on my pillow. I saw him reflected back at me in the way the space had changed: the garden bed, the sharply cut grass, the tidied engine shed. The view of the forest felt different too, as though my perspective had changed, expanded.

Holly sat on the sloping lawn, ever loyal, watching the trailhead and waiting for Jay. Wondering for the two of us where that lovely human had gone and when he would come back.

CHAPTER NINETEEN

I peered through my binoculars at a hazy, viscous substance rolling towards me over the western horizon, one hundred kilometres away. It looked as if someone had smudged charcoal where the sky and land converged. What a strange-looking storm, I thought to myself. The southwesterly winds pushed the dark wall closer until I realized it wasn't rain—it was a heavy curtain of smoke. It looked like a bruise that was just starting to heal: chalky grey and yellow and purplish black. The opaque smoke rolled over the tower and surrounding forest until I couldn't see beyond the bush.

"XMA26, this is XMA567," I said over the radio. "My visibility is down to only one kilometre now."

"Thanks for letting us know, Trina," said Dar, who was on shift that day. "Yeah, it's getting pretty bad out there."

Firefighting crews on scheduled helicopter patrols were getting turned around. The dense smoke made it difficult, if not impossible, for the pilots to see ahead. There was enormous risk for collision with another aircraft. One pilot described it as "flying through toothpaste," while another told me it was akin to running with "a grey pillowcase over your head."

"The ceiling is too low," said a crew leader over the radio. "We're heading back to base."

My eyes stung, my throat felt dry and tight. The forecast was calling for a high chance of lightning today, but I couldn't see beyond the bruise. I had no idea whether the clouds were gathering and growing into towering cumulonimbus. I kept my ear pressed to the sky, listening for the slightest vibration or rumble of thunder.

My southerly neighbour sent a text message:

"There could be lava, or a dinosaur, or Godzilla out there and I'd have no idea."

The heavy band of smoke, a conglomerate of carbon particles, had travelled hundreds of kilometres from British Columbia, where more than a hundred wildfires were still burning out of control.

B.C. had been ablaze for over a month now. The province hadn't yet lifted the state of emergency and tens of thousands of people had been evacuated from their homes. Hundreds of thousands more were affected by the black smoke that had swallowed the skies and hung low, filtering into their communities, homes, lives, and lungs. Environment Canada had issued an air quality advisory in B.C., encouraging vulnerable people—those with medical conditions, the elderly, and infants—to stay indoors. On August 2, Kamloops had the worst air quality rating in the province, reaching an improbable 18 out of 10 on the Air Quality Health Index.

Last week, Jay's unit crew had finally been called to work on the front lines in central B.C.. They were based at a makeshift bush camp with several other crews from B.C.. They had driven through several of the affected communities en route to the wildfires, and were shocked by the dark-violet, apocalyptic sky. Jay texted me a photograph he'd taken at nine o'clock in the morning that I couldn't get out of my mind. The sun had risen hours before, yet the sky was stained black so the street lights glowed orange.

Wildfire smoke is a nomadic phenomenon. Depending on the size and intensity and type of the burn, a body of wildfire smoke can travel upwards of tens of thousands of kilometres.

Seventy years ago, a wildfire that swept right over the current location of my tower had created a similar monstrosity of smoke. On September 22, 1950, the Chinchaga Firestorm, a massive wildfire complex that burned an area of over three million hectares—five times larger than the Horse River Fire of Fort McMurray—raged across northwestern Alberta. It would become the largest wildfire ever documented in North America, and formed a dense plume of smoke so concentrated and voluminous that it became known as the Great Smoke Pall.

The Great Smoke Pall travelled 360 degrees around the globe and contributed to a startling and rare phenomenon: it obscured and changed the colour of the sun and moon to various shades of blue. On September 26, 1950, the sun ascended over Scotland like a sapphire in the sky. Sightings of a blue sun and moon were reported in France, Belgium, Portugal, and Switzerland. People didn't know what to think. There were suspicious reports of nuclear war and religious fears of the Second Coming of Christ. When a blue sun appeared over the western sky, Danish citizens reportedly went to the bank and demanded their life savings so they could flee from the impending disaster.

How far would the wall of smoke from the B.C. wildfires travel? I called Sam to discover that the smoke had reached the far northwestern corner of Alberta, swallowing his tower whole. Sam eventually climbed down and shut himself in the cabin, windows and doors closed, although he had to leave one window slightly open: with the propane gas heaters and stoves, lookouts could be at risk for carbon monoxide poisoning. We remarked on our unavoidable exposure to the harmful carcinogens. Sam brewed tea and opened a book. No point in staying up in the sky and steeping in the smoke.

My friends in Edmonton texted me photos of the city skyline—black-silhouetted skyscrapers rising up against a dusky orange sky. "What the hell is happening?" a friend wrote.

By late in the day, the smoke had become so thick that I had zero visibility from the cupola. Even if the whole forest were on fire, I'd have been none the wiser. The smoke and radiant heat were making my head pound and I felt dizzy. I wished I had a mask, or something to protect my lungs—but no luck. Like Sam, I decided to climb down and seek refuge in the cabin. As I reached for my climbing harness, the phone rang, splitting my thoughts in half.

"Well, have we seen enough today?" said Ralph, irony hanging off his words.

Over the past thirty years, he said, he'd never seen the smoke so bad. Big fires and big smokes. Would these heavy, ominous skies become the new normal in Canada?

"Pray for rain," said Ralph.

More than three million acres of grasslands and forest would burn in B.C. in the summer of 2017. While tens of thousands would be evacuated from their homes in the central and southern interior of the province, the smoke from the wildfires would affect hundreds of thousands more. It was becoming normal for citizens to wake up and, instead of listening to the daily weather report, tune in to the daily smoke report. B.C. pharmacies reported a shortage of surgical masks in the wake of smoke reports that echoed the effects of Beijing's toxic smog. Organizers of the world's steepest race, the Red Bull 400, which treks up the Whistler Olympic Park ski jump, cancelled the event because of poor air quality.

Although it's mostly water vapour and carbon dioxide, wildfire smoke contains a toxic cocktail of compounds, including benzene, which is highly carcinogenic. The most dangerous hazard of smoke, however, is the particulate matter (PM). PM contains microscopic solids, or liquid droplets, that can remain suspended in the atmosphere for extended periods of time. When inhaled, PM can cause serious health problems, especially for those with asthma or other respiratory conditions.

"Stay indoors," urged Environment Canada. "Close your windows and doors."

The dark-grey smoke from the B.C. wildfires would last for days, even weeks. It was difficult to say how far and wide the smoke would travel, or how to measure the negative consequences on human health in communities near and far from the flames. Scientists and health officials knew enough to declare: it wasn't good. Citizens felt, both physically and psychologically, a sense of impending doom or loss. *Ecological grief, anticipatory grief, climate grief*: psychological terms to describe the sorrow related to experienced or anticipated ecological loss. Thanatologist Kriss Kevorkian, who studies bereavement, calls the grief we feel from loss of control over man-made or natural disasters "environmental grief." She coined the term after witnessing the decline of killer whale populations and experiencing a grief so profound she felt as though she was losing members of her family. Many scientists determined that these long summer days of smoke could indeed become the new normal, and that wildfires in Canada's boreal forest, and around the world, would continue to spread faster, burn hotter, and release higher volumes of carcinogenic PM into the atmosphere.

The dark, sinister smoke that cancelled out the sun, moon, clouds, and sky signalled a warning to everyone: welcome to the Pyrocene, the Age of Wildfire.

Weeks later, our prayers were finally answered. The rain came and scrubbed the air clean of carbon and soot.

I climbed my tower, feeling relief at every rung. My eyes travelled up to the azure ceiling of the sky, marvelling at the moving clouds, which cast nets across the forest and reached far towards the distant horizon. The rain had cleared the sky of dust, pollen, and smoke. It was as if someone had taken a rag and wiped a dusty chalkboard clean again.

Up in the sky, I grabbed my binoculars and searched for my westerly neighbour's tower, seventy kilometres away. Only on rare days like today, after a hard rain had swept through the forest, could I locate his tower. *There*. I strained to see it—a black hair rising up along the back of a blue, sleeping giant. Funny that seeing such a small mark, a blip on the horizon—barely a hair—could make me feel so placed in my geography.

"I know where I am," I said to nobody but myself.

It was akin to being a sailor, lost at sea for days or weeks, and seeing land in the distance.

By mid-August, the fire hazard had swung back up to high. Monsoon season in the boreal was over and the forest was once again drying out. The summer was coming to its peak: the raspberries were so over-ripe that they dropped off the branch, the cloudberries grew mushy in the muskeg, and the bilberries and blueberries were fat and plentiful. The highbush cranberries would need another month, along with a hard frost, before ripening neon red on the branch.

I resigned myself to the cupola knowing that the intensity of the fire season was now waning. The forest had received ample rainfall over the past month, and though it was drying out, the speed at which wild-fires would spread had been significantly reduced. If anything lit up, it would be easily contained. Many of the district's firefighting crews had been exported elsewhere—to crises in B.C., to the grasslands and forests of Montana and Oregon. I knew I wouldn't see Eric's or Grant's crews out here again. Man-up days at the tower were long gone.

Jay and his crew were finishing up a long three-week shift on the front lines of one of B.C.'s large wildfires, which was still categorized as out of control. Our communication was limited to text messages and he didn't have cell service, so I received scattered updates. The firefighters were working sixteen-hour days; rain had come to B.C. and yet hot spots from the wildfire continued to smoulder. Jay and his crew trudged with sopping wet boots through the bush, following a gridded map, looking for any lingering hot spots to extinguish. They worked until ten o'clock at night, sunset, before shovelling down hot camp food and crawling into their one-man tents, their bodies cold and wearied. They counted the days until the end of their shift.

"I can't shake off the cold," he said in a text message. "My body misses your body."

I counted down the days on my calendar: another three weeks of solitude. The end of the fire season wasn't so far out of reach. I imagined we were all wearied—lookouts, radio dispatchers, crews, pilots, and duty officers—and ready to leave another season behind.

Firefighting crews and managers in Fort McMurray had finally called EX on the Horse River Fire. Where the wildfire had torn through the bush, green up was already evident amongst the charred standing trees. Life poked through the ashy soil. Forest succession returned to its infant stages, ushering forth pioneer species, including fireweed and willow and young aspen seedlings. The boreal had witnessed and played host to wildfires for thousands of years; the Horse River Fire was just the most recent one to move across the landscape. But inside the city of Fort McMurray it was a different story.

Through my previous work in international development, I'd learned how political conflict and natural disasters can impact the emotional and psychological well-being of communities. The people of Fort McMurray would have difficulty shaking the trauma of what they'd witnessed, endured, and lost. The sudden evacuation, together with the loss of homes, possessions, public buildings, and a sense of security, had shocked people to the core. Some of those who had left the city would never return. And those who remained were forced to rebuild from the ashes, if not literally then psychologically, trying to make sense of what had happened and how the community could best move forward. More Canadian communities will grapple with similar stories in years to come. How do we cope with the effects of wildfire? Where should we build our homes? How do we prepare for or prevent a natural phenomenon? How must we respond when fires threaten the edges of our lives?

As the human footprint expands into the Canadian boreal—residential, industrial, and agricultural development—these questions won't go away; they'll only become more complex.

Many wildfire scientists are advocating for the increased practice of low-intensity, controlled "prescribed burning" on the landscape during the early spring, when conditions are cool and wet. Prescribed burns target areas of grassland or forest that could be at risk of igniting, removing the "fine fuels," or surface vegetation, and thereby reducing the intensity of a potential wildfire. Firefighting crews and wildfire rangers are on site to manage and monitor these controlled burns. Indigenous peoples in northern Alberta, including the Cree, Beaver, Dene, and Métis, traditionally used fire as a tool for ecological and cultural management of land. In the Peace Country, Forestry partners with several Indigenous communities and crews to "fire smart" during the winter months. Indigenous crews clear brush and low-lying branches—"ladder fuels" that wildfire can easily climb from the ground up—and safely set fire to brush piles. Many scientists agree that we ought to heed the wisdom of traditional Indigenous practices, employing fire ecology as a tool for regeneration. In the Age of the Pyrocene, controlled fire can be a solution to prevent the megafires that threaten life—human, plant, and animal.

One long day of nothingness bled into another long day of nothingness.

Time was a beetle moving across the boreal, crawling, inching, scuttling. Units of time had doubled, tripled, quadrupled at the fire tower.

I measured time by the length of the carrots that I harvested from my garden and the height of the sunflowers that stretched towards the blue of sky, their heavy, tightly budded heads drooping low. I checked on them every morning to see if they'd blossomed.

But their flowers remained hidden.

———

I found the words again. Writing wasn't just a tool for survival—it was a way of travelling deeper into my solitude, connecting with nature and the nature of my own thoughts. When I had a pen in my hand or my fingers poised at the keyboard, I belonged to myself.

I began to write about my life at the fire tower. I submitted a two-thousand-word feature to the *Calgary Herald* and they agreed to publish my story and photo essay on the front page of the weekend section. A friend mailed me a copy of the publication and I stared back at myself on the front cover. They'd selected a photo of me gazing through binoculars at the forest below. I'd set the camera on the Fire Finder, using the timer feature, in order to capture the self-portrait. EYES IN THE SKY, read the title, with the subtitle: A LOOKOUT OBSERVER IS LIVING PROOF OF THE OLD SAYING, "WHERE THERE'S SMOKE THERE SHOULD BE EYE-BALLS." TROUBLE IS, THERE ARE ALSO BEARS, BUGS, AND BOREDOM. Here was my story, my private world, made public.

The story offered only a snapshot of life as a lookout, but I received several emails from curious strangers who wanted to know more:

How does your dog get up into the tower?

Have you ever been struck by lightning?

What's your favourite part about the job?

What surprised you the most about the experience?

Don't you get lonely out there?

Don't you go crazy out there?

I could never do that.

I heard from other lookouts, and even a few rangers and firefighters, about how much they'd enjoyed the article. "I don't even like to read," joked one ranger. "But you know how to tell a good story." Soon after, I began to wonder about organizing the stories I'd penned whilst up in the sky and burrowed in my cabin on those long, rainy days into a book. I knew I'd need to open up—make myself vulnerable—to tell the story that wanted to be told.

CHAPTER TWENTY

At the end of a long shift up in the sky, I kicked my feet up onto the cupola window's ledge and flipped a page of my book. The day had slowly drifted by without any smoke sightings. The clouds were plentiful but harmless, piling up like cars bumper to bumper, stuck in traffic.

RRRRRRRRRRRRRRRRRRRRRRR!

My spine snapped upright. What the hell was that? I looked out at the sky, expecting a helicopter to burst into view. But no. That wasn't the whipping sound of rotor blades. That was the sound of an engine!

RRRRRRRRRRR!

I leapt to the window and looked down just as two ATVs came firing out of the bush, whipping doughnuts on my front lawn.

"Woohoooooooooooooo!" yelled a male voice.

I nearly choked on my panicking heart. My worst nightmare had finally come true. That question ringing in my ears: What do you do if two hunters show up at your tower and refuse to leave? I could see Holly standing guard at the bottom of the tower, her fur erect.

"Triiiiiiinnnnnnnnnnnnnaaaaaa!"

And then I looked more closely at the intruders' expressions and realized that they weren't hunters. I knew these people—they were my friends! Four of my friends from Peace River: Luke, Brett, Brianne, and Heather. The same ones I'd hung out with on my final night before the season, listing ridiculous things for me to do at the tower.

"Oh my god, YOU GUYS!" I hollered down at them as they pulled up to the yellow cabin.

I threw on my harness and sped down the ladder. My limbs shook with excitement and my knees buckled the whole way down. Slow down, I cautioned myself; I was going to grab for air, miss a rung, and unintentionally test the fall arrest system.

Heather was already waiting at the bottom, a huge grin splayed on her sun-freckled face, and Holly—having realized the visitors weren't a threat—was jumping happily around her.

"Surprise!!!" Heather exclaimed, smiling ear to ear. "Did we give you a heart attack?"

I shook my head, speechless, incredulous, and hugged my adventurous friend who'd just ridden fifty-some kilometres through the bush, following a labyrinth of trails that were barely trails, thick and practically impenetrable with alders and willows, to reach my fire tower.

"Group hug!" yelled the others, and we crashed together, a mass of limbs and laughter.

My friends were well equipped to undertake such an adventure into the bush. Luke was a wildlife biologist; he had even worked a few years with my father in Peace River before he retired. Heather grew up on a bison farm. As a university student she'd spent six summers working at a remote hike-in lodge in Assiniboine Mountain. Brianne was a documentary filmmaker, who'd also grown up in the Peace Country. She'd just wrapped up a film project about bushcraft, the art of survival in the woods. She filmed herself camping out in the woods, alone, and taking a week-long canoe trip on the Peace River. Brett was a wildfire ranger in Peace River. He had studied forestry and resource management in

New Brunswick and had worked a number of seasons as a wildland firefighter in the Slave Lake region.

The group was sweaty and exhausted and bursting with stories about the seven-hour journey into the fire tower. The quads had nearly sunk deep into the muskeg as they manoeuvred around a beaver dam. The chainsaw they'd brought with them didn't work. At one point they nearly turned around. And the whole way here, they weren't entirely sure if they were on the right trail.

"I knew we were getting closer when we hit the hills," said Brett, laughing. "And then fifty metres away I saw the sign for your helipad and I knew we'd made it! That's when I gunned it!"

I was so enthralled by my friends' presence that I nearly missed dispatch calling me on the radio.

"XMA567, you'll be on high tomorrow," said the dispatcher, doing the evening radio rounds.

"That's all copied," I said, trying to squash down the elation in my voice and sound as if it was just another day at the fire tower.

My friends had packed in enough food to stay for a week, although they'd only stay for the long weekend. They'd brought all their camping gear, prepared to camp out near my firepit. Heather, the planner of the group, had mapped out our meals: roast chicken and potatoes, vegetable soup, bacon and eggs, sandwiches—the list went on. I practically drooled as she listed off the weekend menu.

Later, we gathered around the campfire, roasted hot dogs, drank hot chocolate, stuffed our greasy faces, and talked and laughed. We laid down our sleeping bags, convincing ourselves we'd sleep beneath the stars. Backs pressed against the earth, we talked until the sky drained of light. Holly sat and kept guard, smelling the wild creatures lurking beyond the edge of the bush, drawn in by the scent of our campfire and food and warm bodies.

"Hey, look!" cried Heather, pointing over the tops of the spruce.

The northern lights were faint, hazy against the darkening sky, barely discernible. We watched for over an hour as they deepened into a soft, ghostly-green shade. They moved like a dancer's silken scarf. The mosquitoes from the muskeg descended upon our exposed faces and limbs, forcing us to mummify ourselves in our sleeping bags.

Eventually, we succumbed to the sounds of the forest and mosquitoes droning overhead. I felt safe and happy and calm as, one by one, we fell asleep. And then, one by one, we silently rose, tormented by the bugs, seeking refuge inside the cabin.

When I rose at 7 a.m., readying myself for morning weather, I could hardly believe my eyes:

Four humans curled up in sleeping bags and one dog, dreaming deeply on my cabin floor.

How many days had I sat up in the sky, watching an exquisite show put on by Mother Nature—an explosive sunset, a ghostly fog rising in the forest after a heavy rain, a red-tailed hawk riding the ridges from north to south—and wished I could share the beauty of the view with someone? And suddenly I had four new sets of eyes to see the world with.

"Wow," said Heather breathlessly. "The forest seems to go on forever."

She stayed up in the cupola until sundown, long after I'd climbed down from my shift, scouring the horizon with astonished eyes. She seemed to feel what I felt when looking out from the fire tower: freedom for thoughts to roam, contemplate, and meditate. A sense of home.

"I think I see a smoke!" Heather cried, gazing over the northwestern horizon through my binoculars. She paused. "No," she said finally. "It's just the way the sun is lighting up a cutblock."

They stayed four days at the tower, delaying their departure by a day owing to a rain that drenched the forest and drowned out the trails. We barely left that small clearing in the woods. We cooked delicious meals, drained my coffee rations, lounged in the cabin and outside in the sun, picked wild berries, gathered around the campfire, and set off fireworks beneath the tower. Heather came up with a scavenger hunt and the five of us spread out on the land, looking for wild things—a pine cone, a frog, a caterpillar, a green rock, and a blueberry. We ate ants from an anthill on the airstrip, and I felt as though I'd been transported back to Uganda. The ants were crunchy, though their taste was reminiscent of lemons. We slept on the cabin floor like a pack of contented wolves.

Brianne wrote three pages in my tower guest book. Later, Luke would tell me that those four days at the fire tower had been the highlight of his summer. Heather was misty-eyed on the morning they left, though we'd see each other in a few weeks. I readied my heart for their departure as they filled up the gas tanks, strapped their bags to the backs of the ATVs, and hugged me goodbye. I was now used to being the one left behind. But I didn't feel lonely when they drove away so much as grateful. These were friends who had dared to navigate beyond the maps and risked getting lost to find me—the same friends who'd helped me split and stack firewood in the dead of winter.

They reminded me that even when you want to hide away from the world, there are always ways to be found again—to belong to something larger than yourself.

Holly bounded ahead of me and veered off the path, her olfactory system hooked on a moose trail. She rolled in a small shell of flattened grass, a bed for a moose, or even a bear. Two years of hiking together and we'd learned to move separately but together. She never strayed far.

Only one week remained in the fire season.

Soon we'd fly back to civilization. Other than spending some time with my brother's family and meeting my new niece, Brielle, I hadn't yet mapped out the next step, where I'd go, what I would do for the winter. *Women Who Dig* was soon going to be released into the world and I wasn't sure I was ready to be an author; I had more questions than answers, more doubt than certainty. But the thought of standing up onstage at a launch event and reading passages of the book aloud to my friends and family made my heart soar.

Jay and I continued to send text messages back and forth. We spoke on the phone during the evenings, anticipating my release from the fire tower so we could see one another in person again. We had been dreaming of driving across Canada together, bringing Holly and camping along the way. He'd built a small, lightweight teardrop trailer, which he would tow behind his car. Jay wanted to reach southern Ontario—nearly four thousand kilometres away—by the time the maple leaves ripened into brilliant shades of burgundy and scarlet red. I was starving for companionship and movement, and I knew I didn't want to spend another winter splitting wood and scrambling to heat a ramshackle trailer on the edge of a cliff.

"Do you want to be my girlfriend?" Jay asked over the phone.

"Yes," I said. Then I laughed because the question made me feel like a teenager.

I was ready to return to civilization, though I already felt nostalgic for the yellow cabin and tangled, abundant garden, and my bird's-eye view of the boreal. I no longer woke up feeling fear, or anxiety, or the need to distract myself. Today a pilot had dropped off some mail for me and I hadn't even climbed down to say hello. I waved from the tower and watched him fly away as quickly as he'd come.

I belonged here. I belonged to the fire tower, to the forest, and to the black-and-white dog that zipped back and forth, chasing scent trails.

I walked towards Holly then stopped dead in my tracks. She had come to an abrupt halt, facing a stand of thick, silvery-leaved alders. Her whole body was on guard: ears perked, nostrils twitching back and forth. The fur on her neck rose up suddenly, her way of telling me that something or someone was out there, beyond the bush, watching us.

My hand slowly reached for the canister of pepper spray on my belt. It was probably a bear, I thought, feasting on the alder leaves and the carpet of sweet clover. I stood up on my toes, trying to see above the alders. Perhaps it was a moose and her calf. But I couldn't see anything.

And then—so suddenly, so silently—a huge black dog stepped out from behind the curtain of bush. The black dog stopped only a few metres away and stared back at us with intense yellow eyes.

That's a big dog, I thought. No. *Wait.* My brain struggled to process what I was seeing.

It was a wolf, I realized. A black wolf standing only a few strides away from me.

It's beautiful, I thought, rapidly, illogically. It was so terrifyingly beautiful that I couldn't breathe.

The black wolf was double the size of Holly. Those legs! The animal stood tall, much taller than I'd thought a wolf could be. I'd never been afraid to cross paths with a coyote before, but I felt vulnerable, frozen in the wolf's presence. For a fraction of a second I forgot that I was even human. I tumbled down into those yellow eyes, distant planets. *I know something you don't know.*

I wasn't afraid. I knew there were very few accounts of wolves attacking humans. But then I glanced down at Holly, and she looked small and fragile compared with her wild canine counterpart. She didn't bark or growl, but instead stood mutely by my side and stared back at the black wolf.

The wolf wasn't afraid of us either. It wasn't backing off, slinking into the bush. It could be hunting Holly, I thought. I had heard stories of wolves going after dogs, ambush attacks led by an alpha wolf. I did a quick 360 scan around us, worrying if this was the moment when four other wolves would slip out of the bush, the pack surrounding us.

Holly looked up at me with her large, amber-flecked eyes and wagged her tail nervously, and I felt an enormous wave of maternal love, and a rising fear. Then I reacted without thought, firing a quick shot of the pepper spray. A violent orange plume burst forward and the black wolf leapt back, startled, then turned towards us again. *Those eyes.* Staring yellow.

"GET OUT OF HERE!"

I could barely recognize my own voice. Adrenalin coursed through my veins.

"GET OUT OF HERE!" I screamed again, with all my breath.

The black wolf spun and loped back into the forest. My heart pounded and I grounded a trembling hand on Holly's back. She looked up at me curiously, awaiting my next move.

"Let's go home, girl."

Who are we without our shadows? What do we know about ourselves until we come face to face with fear? I realized that, if that wolf had attacked, I would've defended Holly with my whole body. I'd kick and scream and do whatever it took to drive that wolf back into the bush.

I would defend Holly as I'd defend myself.

Later that evening, I opened my mail. There were a few letters from friends and a package from Sam, my tower neighbour in the Far North. I ripped open the package and held a book in my hands. I felt something electric run up my spine and my whole scalp tingled.

"No way," I said. "You've got to be kidding me!"

Sam had sent me a worn paperback copy of *Women Who Run With the Wolves*.

CHAPTER TWENTY-ONE

"**I**'m not coming back next season," Sam announced.

"Oh," I said, my heart falling, unable to conceal my disappointment.

I looked north, pressing the phone to my cheek, improbably searching for his fire tower on the horizon, though I knew I'd never see him there, five hundred and some kilometres to the north. Was he doing the same? Looking south to place himself on the map of our friendship?

Sam was flying back to Halifax tomorrow, closing up his fire tower for the winter. He was eager to leave, though sad to say goodbye to the shrunken forest of scraggly black spruce, the acidic soils where he grew kale and lettuce for both himself and a colony of rabbits. Would he ever come back? Probably not, he knew. But he was ready to reintegrate into society. Sam dreamt of wandering down to California to study permaculture and meditation, and eventually starting his own social enterprise. He wanted to buy a piece of farmland outside Halifax and host community, arts, and activism workshops and gatherings.

He wanted to be a part of the wider world again.

Sam shared with me something that he'd recently written, an essay about longing and belonging, about contemplating the transition from solitude back into civilization. He wrote about his sense of excitement for human contact, but also his fear of how solitude had, perhaps, made him fragile. How we live in a world that doesn't often value vulnerability or emotion.

> I don't know how I'm going to handle learning about where the world has gone. I'm not sure if I can handle the game of pretending. Where one hand holds one mask and the other searches frantically for a replacement. Where we pretend that we are anything but wondering souls, groping in the dark. Where we pretend to not want or need love.

I could relate deeply to Sam's words. I didn't want to fall back into old patterns of denying my emotions, of not facing my own truth, of distracting myself and looking away from the smoke. The tower had given us both the opportunity for space, reflection, and connection. And the most unusual kind of friendship. Even knowing that a part of him would always dream about nature and solitude at the fire tower, Sam was ready to root himself more firmly in Halifax. He was ready to explore, once again, what scared him most: the messiness of human relationships.

Although I was happy for my friend, I missed him already. Sam had become something of my true north. A friend to call on the hardest, loneliest days—and the joyful days, too. If I returned, it would be peculiar to be in my cupola without the knowledge that he was up in his own tower, or working in his garden, or wandering the forest.

But he wasn't the only one not coming back next year.

After over three decades of watching for wildfires on the horizon, Ralph was finally retiring. These waning days in the forest would be his last days as a lookout.

"I guess it's time to throw in my hat," he said tenderly.

Ralph had spent thirty summers watching weather and wildfire. Thirty years of feeling nature strip the soul bare, expose hard truths, burn the forest down, and start anew. I could feel his enormous sense of loss, and imagined how strange it would be for him to drive away from his fire tower for good in late September.

"What am I going to do after?" Ralph asked playfully. "I'll be at the Chinese restaurant in town and watch a storm roll by. I won't be able to stop watching for lightning. 'First strike!' I'll holler, and probably give the poor waitress a heart attack."

You can take a man out of the fire tower, but you can't take the fire tower out of a man.

Over the past century, societies have been slowly decommissioning fire towers. Most of the world's fire towers—lookouts atop mountains, wooden towers, steel towers, and trees with platforms—have been abandoned by people, left for nature to erode, rust, or swallow whole. Some have been preserved for historic purposes, or tourism.

Very few fire towers are in active operation today, although some do remain in Australia, Latvia, Mexico, New Zealand, the United States, Poland, Brazil, South Africa, and Turkey.

The province of British Columbia began decommissioning fire towers in the 1970s and throughout the 1980s because of budget restrictions, and only a few functioning towers are left in the province today. Some wildfire managers insisted that towers weren't the most effective means of detecting wildfires; they claimed that aerial patrols, and satellite and lightning detection technology, are the way of the future. B.C. relies on citizen calls to report many of their wildfires. Satellite imagery detects wildfires in remote areas of forest, although often the blazes don't show up on the radar until they've grown in size and intensity to at least three hectares. Lookouts in Alberta are required to catch smokes at 0.01 hectares. I wondered about the massive wildfires that ignited and spread in 2017; would some of them have been detected faster by a lookout?

In 2013, Saskatchewan closed its last fire tower. Cameras would replace lookouts, the government asserted. One government official claimed that the primary issue was safety, as lookouts were free-climbing the eighty-foot towers—never mind that they could have installed fall arrest systems. Would early detection from lookouts have made a difference in 2015 when wildfires blew up in northern Saskatchewan, forcing thirteen thousand people to leave their homes?

Trained human eyes and sound judgment of wildfires will always be superior to cameras in detecting and categorizing smoke. A wildfire would need to build into a considerable size before the smoke would be visible on digital cameras—adding to the risk that the fire could blow up into a campaign fire, or megafire, and cost governments tens of millions of dollars to manage—but an experienced lookout was more likely to "catch 'em small," as we liked to say. There are major cost advantages to managing fire towers for detections over the frequently cited alternatives. Helicopter patrols are an exorbitant cost—in terms of both dollars and environmental impact. Camera installation and maintenance at remote tower sites would involve increased reliance on helicopter flights, jet fuel, and technicians. Wildfire scientists agree that ignition and spread of wildfires is more volatile than ever. Megafires are costly to manage, often requiring tens of millions of dollars of resources to fully extinguish. Lookouts may seem archaic to the public and politicians, especially to those who tout "modern technology" as the solution to every perceived crisis. But as my tower neighbour loved to say, "Don't fix it if it ain't broke."

Fire towers remain an important lifeline for wildfire detection in Australia. The state of Victoria has seventy-four fire towers strategically located throughout the forests. A lot is at stake in Australia, where wildfires—or bushfires—can move at deadly speeds. On February 7, 2009—a day that became known as Black Saturday—more than 400 bushfires broke out, eventually killing 173 people. A decade later, in June 2019, Australia's bushfire season ignited in the southeastern part of the country. Six months later, the fire season was still raging. By

January 2020, the bushfires in Australia had burned over 18 million hectares—46 million acres—killing at least thirty-four people and destroying nearly six thousand buildings, half of those homes. Scientists estimated that over a billion animals were killed in the rapidly moving fires, and some endangered species may have been driven to extinction. The smoke from the bushfires travelled over eleven thousand kilometres across the South Pacific Ocean to Chile and Argentina. In New South Wales and Victoria, multiple states of emergency were put in place to evacuate tens of thousands of Australians from their homes.

In the Age of the Pyrocene, the early detection of wildfires and bushfires has never been more important. The lookout's unofficial mantra of "catching 'em small" isn't about pride or nostalgia; it's about containing and extinguishing wildfires as early as possible—during peak burning conditions—in order to avoid the explosion of large-scale wildfires that threaten life.

Alberta has the highest concentration of fire towers in Canada, maybe in North America, and quite possibly the world. I'd argue we have one of the fastest, most efficient wildfire detection programmes on the planet.

"You can never replace a set of knowing eyes on the landscape," a ranger once told me. "We need all facets of detection to manage wildfires as quickly as possible," said another, "and lookouts play a major role in that." Indeed. Statistics show that lookouts catch nearly 50 percent of all wildfires, in every fire season. How many of those fires would've blown up had they been detected, say, a few hours or days later?

Sadly, the decommissioning of fire towers is becoming a common story around the world. Some people are even capitalizing on these abandoned towers. In the U.S., former lookout towers are being turned into tourist destinations you can book online. *Stunning Fire Tower Vacation Rentals!* The U.S. Forest Service rents out over sixty fire towers in the American west. *Spend your next vacation spying storms and stars!*

I cringed at the flashy advertisements. The idea of my tower being converted into a tourist scheme for couples looking for an exotic weekend getaway felt like an insult. The fire tower represented one of the last jobs of a pre-automated world. Most lookouts would tell you: it's sacred territory. It's more than just a job—it's a home.

What will be lost if the last tower shuts down?

And, more importantly, how much will it cost us?

"What keeps you coming back season after season?" I once asked Ralph.

"I love watching how the land changes every year," he mused. "How the trees grow up and burn down. I observe how the mother grouse protects her chicks from the hawk, or how the wolves take down a moose. It's neither right nor wrong—it just is."

What keeps lookouts coming back season after season? A love for observing nature. Solitude. And a singularity of purpose: detecting wildfires while they're small and thereby preventing damage to people, forests, and wildlife.

A storm was brewing to the southwest of my tower, but my expectations for wildfires were low. The intensity of lightning in late August was minute in comparison with the violent storms of June and July. But this storm had conviction. The cloud doubled and kept rolling towards me, so I closed up the cupola windows.

The winds churned and I heard the violent whirring of the anemometer. The cupola began to rattle back and forth atop the tower. The storm was going to come right overhead, I realized. The trees flattened and my laundry, hung on a clothesline below, went flying in every direction, a tempest of colourful socks, underwear, and T-shirts. Holly slunk beneath the cabin porch as the storm drew nearer.

Rain lashed the windows and the eye of the storm seized the tower. My water bottle went crashing off the Fire Finder onto the floor as the cupola shook. The sky began to hurl hail, the white ice ricocheting off the windows like gunfire. I jumped on the radio to report the weather shift.

"26, THIS IS 567 WITH A WEATHER UPDATE!"

I couldn't even hear the dispatcher above the hail, but the radio snapped, so I kept on.

"I'M GETTING HAIL—ALPHA ONE-ONE! WINDS ARE MORE THAN NINETY KILOMETRES!"

Grape-sized hail pelted against the windows. I put down the radio.

"HOLY SHIIIIIIIIIIIIIIIIIIIIIT!" I screamed into the abyss of the storm.

Minutes later, the storm had pushed off to the east and the rain relaxed to a drizzle. The ground sparkled like white granite and hail piled up against the yellow cabin. I cringed to witness the devastation of my garden: My peas, flattened. The lettuce and kale tattered. Holly emerged from beneath the porch. She looked around, shook out her damp fur, and sniffed at the hail.

A colony of ghosts proliferated in the distance, spooky grey columns of moisture rising from the forest. No smokes. Suddenly, I spied a stab of white light—*dry lightning!* The cloud opened up and the rain formed a solid curtain of blue. I looked away then glanced back, and there it was: a column of white smoke, pushed low to the treetops with the surging winds. Why did it always feel like a small miracle—seeing smoke? A kind of magic trick: now you see it, now you don't.

I radioed in the smoke location, though the smoke quickly disappeared, doused by rain.

"No smoke in sight," said a crew leader. "But we're seeing a lot of spooks out here."

"It was definitely *not* a spook," I replied, although not over the radio.

"XMA567, this is Yankee Whisky Bravo on 202."

I reached for my mic. "Go for 567."

"Good afternoon, we're just coming up on your detection," said the crew leader. "We're not seeing anything. What was your bearing on the smoke? We'll fly the bearing from your tower."

I passed along my bearing and detection and watched them through my binoculars, the helicopter a black fly circling the location of the phantom smoke. I was confident I'd seen a wildfire and figured it must be hidden, barely burning somewhere below the treeline.

"We've had a good look around, but nothing found," said the crew leader to the dispatcher.

Away they flew. My heart sank.

"It was a smoke," I said to nobody.

Stubbornly, I kept watching the blue plateau over the western horizon. Maybe my distance had been off, I thought. Maybe it was farther than I thought across the flat stretch of forest. I stayed vigilant.

Suddenly, a column of white popped back up. I grabbed the radio to inform dispatch and listened to my westerly neighbour call in a cross-shot on the smoke. I scribbled down his bearing, then went to the cross-shot map to pull the sewing thread from his tower to mine, fixing their intersection with a tack. *Aha!* A precise location of the smoke. My distance had been off by ten kilometres, but what did it matter? They would now see the smoke rising out of the forest.

"Confirmed wildfire," said the crew leader. "What's the next fire number, please?"

It was a small fire, only 0.2 hectares, and burning a path towards a meandering creek. It wouldn't take the crew long to get it under control. I watched the smouldering drama through my binoculars and felt an odd sense of ownership. Finally, I thought, I'd trusted my gut. I hadn't backed off my judgment or doubted what I thought to be true, even when the crew came up empty-handed. I'd kept watch, patiently, willing the wildfire to the surface with my eyes.

What was it about being the first to observe nature in action? Witnessing what the forest had done for millennia? It made me feel small and insignificant, and yet also like a quiet hero. I was proud to be a small part of the collective effort to spot the beginnings of wildfire, joining the legacy of the women and men who watched before me. It was lonely and thankless work, and in the moment of reporting a smoke there was no one with whom I could share the glory—no one but myself. But it didn't make me sad; instead, I felt powerful and connected and deeply useful. The smoke would fade into nothingness, the wildfire into a black scar on the landscape that few would ever fly over. Yet I'd never forget this moment.

Looking out across the blue plateau of forest and sky, my chest ached. I felt the possibility of open space, a canvas for imagining a future in which I could be at peace with myself—and even let go.

I began sweeping out the cabin and packing up my belongings, dreaming of the comforts of civilization. In truth, I was dying to take a long, hot shower—with actual water pressure. I was craving an extra-large, all-dressed pizza with lots of cheese. Wouldn't it be novel to walk into a room and turn on a light switch? Mostly, I was ready for the freedom of new horizons, the possibility of getting in my car and driving anywhere I pleased, maybe even right across Canada with Jay. In less than six days I would no longer be "XMA567"—I'd answer only to Trina. I wouldn't have to report the weather twice a day, nor worry about a fire igniting in my forty-kilometre radius of forest.

Heather and my friends were already making plans for my return to civilization: a Mexican-inspired potluck at Heather's place. I was craving fish tacos and tangy guacamole, but mostly the physical camaraderie of my dear friends.

"I can't wait to get out," I said to Jay one evening.

"It sounds like you're in prison or something," he replied with a laugh.

I cursed the calendar and the clock. Six days . . . five days . . . four days . . .

I was packing my extra dried food rations into a plastic bin one evening when the phone rang. It was one of the wildfire managers, I assumed, calling to inform me what time the helicopter would arrive in a few days to close up the fire tower.

"Actually, we're going to keep your tower open for another two weeks," he said.

My stomach knotted up. He had to be joking. Fourteen. More. Fucking. Days. It felt as though I'd been denied parole. Sure, it was only two weeks. But an extra two weeks, after four months alone, seemed impossibly, painfully long.

"The fire indices are climbing," he explained sympathetically. "The temperature is supposed to get into the high twenties next week. We really need you to stay longer out there. Do you have enough food and water?"

My exhausted body and mind wanted to collapse on the floor and throw a tantrum. *LET ME OUT OF HERE!* I wanted to wail.

"I have enough water," I sighed into the phone receiver. "I'll need some groceries, I suppose."

The season stretched into September.

The weight of isolation bore down on me. I moved about with a kind of resignation and turned to nature for guidance. My hands no longer searched for tasks, for distraction. Instead, I sat for long hours doing nothing but looking out the window, studying the clouds, studying a blade of grass. Green aspen leaves began to yellow. I watched leaves, one by one, let go and surrender to the changing season.

The temperatures hovered at only 5 or 6 degrees Celsius in the early morning and late evening, coating the forest with dew that clung to everything, even the spiders' webs, spun between blades of grass. The fuchsia fireweed went to seed and its leaves ignited with the colours of its fiery namesake. The birch leaves turned bright tangerine.

"I feel like I'm slowly dying," said a lookout neighbour to the north. "I don't know how else to describe it—this level of intensity of being alone, emotionally and physically."

I was moved to tears watching a sunset over the western ridge and the explosion of electric orange-and-pink light. A large black raven landed on the sloping yard and spent hours feeding on blueberries, and I wondered if my grandfather's spirit had come to say hello. I stumbled upon an enormous spiderweb that stretched between the continents of two black spruce trees in the muskeg, and I spent an hour photographing the intricately spun map.

My garden was expiring, yellowed and shrivelled. Minuscule seed-eating birds preyed on the beans and the peas that the hail had ruined. I inspected the sunflowers faithfully every morning, hopeful of spying a peek of yellow, but their stubborn buds remained tightly closed.

"What's your favourite flower?" Jay had asked me in one of his letters.

I thought of the sunflowers Akello and I had planted in Uganda. We sowed them along the edges of the garden and planted them amongst our favourite foods: corn and beans and squash. The sunflowers shot out of the earth, lured skyward by heat and rain, their heavy, seeded heads drooping and obediently following the course of the sun. We took our engagement photos in the garden. I'd worn a white cotton sundress and held a bouquet of sunflowers and purple kale.

I thought about the story of the lookout who had dug two graves beneath his tower. "What did you bury?" they asked him. "My soul," he told the bewildered visitors. When I heard the story, I thought he'd gone crazy, but I finally understood.

I'd come to the tower to let go of my shame and the stories I told myself that were no longer true: that I was bad, foolish, weak, and selfish, that I deserved to be exiled from the world. I'd come back to the fire tower to find the parts of myself that had got lost.

On a perfectly clear day, when rain had washed the dust and smoke clear from the horizon, I could see my reflection on the cupola window: a wild-haired, wild-eyed woman with bruised legs and aching arms and a bursting heart and wide-open eyes. A reflection of myself that I'd grown to embrace, even love.

They say that when you get lost in the woods, you should hunker down, keep your heart rate low. Whatever you do, don't panic. Listen to the forest. With a calm heart, figure out your next move. Seek shelter. Start a fire. Hydrate. Light a smoke signal so others can find you. Or, retrace your footsteps back home.

Getting lost was inevitable. Surely, I'd lose my way again. But now I knew, I'd be okay.

Early one morning, I went walking. Dew clung to the bright-scarlet leaves of the highbush cranberries. I propped up an old metal stool in the middle of the airstrip. I placed a tin can on top and paced eight metres back. Holly trotted behind me, obedient, her carefree demeanour suddenly made serious by the shotgun slung around my shoulder. She'd probably grown up with guns. Many people on the First Nations reserve where she was born had hunted, and she'd been around guns on the farm too. Justin had said she was the smartest hunting dog on the farm. She stayed by my side, watching me with those big brown eyes of hers.

I was comfortable shooting, and not just because of my dad's lessons before my first summer at the tower. When I was a girl, I used to go target shooting and grouse hunting with my dad and brother, carrying my nana's old Remington .22 rifle. Dad wanted me to inherit her old rifle and he taught me how to line up the shot. We learned how to read the signs of grouse, scouting for loose feathers, and observing "dust baths," shallow depressions in sandy soils where the birds burrow in dust as a means to rid themselves of lice and mites. Dad showed us how to field dress the bird, removing the feathers, feet, and head. Those early years of learning how to read the bush, of hunting with my dad and brother, filtered into my life at the tower every day.

I palmed a shotgun shell into the magazine and pumped the action, then shifted my legs into warrior stance and lifted the heavy gun, the barrel resting against my cheek. I flicked off the safety and squinted down the barrel, lining up my shot. My dad's words returned to me: *fire on the exhale*. I took a deep breath, put my finger on the trigger, and
BOOM!
The tin can went flying.
A blast of satisfaction. My eye was good, I thought, my shot was good.
I looked down at Holly, who wagged her tail nervously.

"Did you see that, girl?"

I set up the torn can and pulled the trigger again, pacing back twelve metres, fifteen, eighteen. I pulled the trigger until I felt more powerful than the buck of the gun and the bite of a shot. Until most of the fear had left my body and my breath was calm. I fired until the gun was just a gun, just another tool, no different from a shovel, or a weed whacker, or a Swede saw.

As we were walking back to the cabin, Holly bounded ahead into the tall grass.

SWOOSH!

I jumped, spooked by the suddenness of a spruce grouse flapping frantically out of the grass. The bird flew up to a high branch of a pine tree. Holly looked up at the grouse eagerly, her tail wagging, then back at me.

"Good girl," I whispered.

I pushed a shell into the magazine and quietly pumped the action, then lifted the gun to the sky. The bird was frozen on the branch. We locked eyes. Deep breath: I pulled the trigger, a shot fired out, and the bird dropped soundlessly to the earth.

"Oh my god."

I felt a rush of excitement and satisfaction. A tender kind of responsibility, and even sorrow. My heart pounded in my chest, my hands trembled. Shooting the grouse made me feel both animal and human— intensely capable, and yet surprisingly culpable.

Holly leapt towards the fallen bird. It was a clean shot. I picked up the grouse's leathered feet and spread the brown-and-black-and-white-checkered wings. The feathers were so soft. I'd use them to make a mobile to hang in the cabin. I'd try to use every part of the grouse.

"Thank you," I whispered.

I cupped the bird's head in my hand, feeling its fragility in my palm, then twisted the head. It pulled off as easily as plucking a weed from moist soil. I stretched the wings wide and placed the bird on its back on the ground, then stepped onto the outstretched wings and pulled the feet in an upwards motion, feeling the tear of muscles and flesh as its exterior pulled off like a sweater. I plucked the remaining tail feathers and placed the drooping internal organs and intestines into a pail. Then I detached the wings from the breast meat. A small and humble meal for one.

I reached for the bird's small red heart and held it in my palm. The organ was warm. I gazed down upon the wild heart with awe.

Then I lowered my hand and Holly obliged, tenderly eating the grouse's heart out of my hand.

I thought often about Stephanie Stewart, the lookout who'd gone missing from her tower at the end of the fire season in 2006. Although the RCMP strongly suspected homicide, Stephanie's case remained unsolved—her body has never been discovered. Although I never knew Stephanie, I'd come to know what it meant to be a lookout. She'd climbed the fire tower for nearly twenty years. She'd endured extreme weather, isolation, and even a close encounter with a grizzly bear. I thought about everything I'd encountered in my two short seasons as a lookout, and the fears I'd had to overcome. I wished I could've known Stephanie.

She must have been brave, I thought. She must have been resilient, resourceful, and strong in her solitude. She must have known who she was. I grieved for Stephanie's disappearance and her family, and I prayed that, one day, justice would be served. But also I wanted to honour her bravery as a woman—and as a lookout. I wanted to remember the way she'd climbed the tower, season after season, watching for smoke and helping to keep communities safe from wildfires. It was the stories about Stephanie's commitment as a lookout and her courage to over-come adversity that I wanted to hold closest to my heart.

The night before the harvest moon, I dreamt of the huge grizzly bear bar-relling out of the bush, rushing towards me. Every golden hair lit up in the sun. Clumps of earth went flying. Fear seized my body. I was holding a wooden spear, crudely carved from a fallen tree limb. It was a futile effort, a child's imaginary weapon, and death by mauling was inevitable. I readied the spear and prepared for the blow, hoping the bear would knock me unconscious so I wouldn't be able to feel the pain of his jaws crushing my skull or seizing the back of my neck. Only a moment before impact, I made up my mind: surrender, let go. I dropped the spear.

The bear charged and I felt death brush by with its grizzled tips, but the pain didn't come.

Waking was like a gunshot, a startled field of geese.

The pilot and ranger helped me pack up my few material belongings, and Holly leapt obligingly into the helicopter. I covered the windows of the cabin with plywood. Over the winter the orange-and-black tortoiseshell butterflies would crawl behind the shutters and huddle together for warmth. Some, miraculously, would survive the frigid temperatures with their tissue paper wings. Next spring I'd open the shutters and they'd cascade to the earth like confetti.

Next year, I knew in my heart, I'd be back again.

I made one last climb up to the cupola and covered the Fire Finder with canvas. I cast my wild eyes out into the forest and took a long drink from the blue horizon. Golden leaves fluttered to the earth. I locked up the cupola, my office in the sky, and descended the ladder.

The pilot and ranger boarded the helicopter.

"Wait a second! I forgot something!" I yelled, running back to the garden.

The sunflowers now stood over two metres tall. I looked hopefully beneath their heavy, bowed heads. *Aha!* A single flower had opened. The soft yellow petals had unfurled and one black eye stared back at me. I rushed back to the helicopter, holding a bouquet of one sunflower, and climbed up into the back seat, beside Holly's kennel. *Thump, thump,*

thump. Her white-tipped tail beat out that now-familiar song, a way of saying to me, I'll follow you anywhere.

"Here we go, girl," I said.

The helicopter blades spun, then thundered.

FIRESTORM

I think it would be well, and proper, and obedient, and pure, to grasp your one necessity and not let it go, to dangle from it limp wherever it takes you. Then even death, where you're going no matter how you live, cannot you part. Seize it and let it seize you up aloft even, till your eyes burn out and drop; let your musky flesh fall off in shreds, and let your very bones unhinge and scatter, loosened over fields, over fields and woods, lightly, thoughtless, from any height at all, from as high as eagles.

—ANNIE DILLARD, *Teaching a Stone to Talk*

I wake at five o'clock in the morning. The cabin is on fire. Pulsing orange light halos everything around me: Holly, curled like a snail at the foot of my bed, the single dresser, the wide-open bedroom door, the linoleum cabin floor. The world glows like an ember. I snap out of bed, disoriented by the absolute quiet. Holly jerks awake, gazes at me with big blinking eyes. What are you doing, Human?

It's my first morning of the fire season. My fourth fire season.

The cabin is not on fire—not literally, anyway. But the sun summits the northeastern horizon and the honeyed amber light pours in through the windows. Barefoot, I rush outside to capture the shot: a cloudless sky on fire, pulsing sediment layers of gold, bronze, and brilliant yellow.

It's only early April and the forest is semi-frozen, transitioning from winter's sleep to spring's thaw. I've been stationed at a new tower in the Peace Country, located fifty kilometres east of where I lived for three summers. A settlement tower, only ten kilometres from a highway and less than fifteen kilometres from a Métis farming community. The Peace River watershed, meandering north, is only twenty kilometres to the east. The last lookout spent ten seasons here before opting to move to a tower in the southern part of the district.

I photograph the brilliant sky, trying to preserve the colours, though I know from experience my efforts are futile. The exquisite glow won't be contained in a photograph.

I consider the irony: I'm not supposed to even *be here*.

When I flew away from the fire tower in 2018, after my third season of working as a lookout, I promised myself it would be the last. Not because I'd had a hard or bad season. Quite the opposite: my third season felt like a coming home—to the forest, the profession, to myself. I caught seven smokes that season. I'd experienced what it is like to be up in the cupola when the tower is struck by a bolt of lightning. *Flash! BOOM!* Every hair on my body shocked awake.

Most significantly, I penned out the first draft of my second book— a book about the fire tower—and came to realize that I wasn't writing environmental journalism, a book that explored hard truths about wildfire and climate change, as I'd initially intended. I was writing a memoir. I was finally summoning the courage to write about my personal story, what forces led me back to the North to write from my perch one hundred feet up in the sky.

When rain lashed the cabin for a week straight, I no longer felt trapped, or stir-crazy. I let the fog drape over the tower like a woollen blanket. I stayed in bed and my fingers flew across the keyboard. A ranger who came to drop off groceries gawked at my cabin walls, which had morphed into rainbows with colour-coded sticky notes that outlined my book and its elements. I wrote on high hazard days from the sky amongst the low-hanging clouds and the acrobatic tree swallows and American kestrels that perched atop the cupola. I was no longer afraid I'd miss a smoke. On the contrary, I felt as though I could predict them, based on the knowledge I'd accumulated about the forest and the buildup of weather and how fires are born in the boreal.

I joked that I could wake up and step outside and feel in my bones if something would burn today. It was important to be prepared for the unforeseeable, but I'd also developed something like a sixth sense, an instinct for the days when the forest lit up.

At the end of my third season, I felt as though I'd come full circle. The fire tower had challenged me, shaped me, grounded me. Maybe it was time to reintegrate more permanently into mainstream culture and society, as Sam had done last summer. He still longed for the tower he'd never return to, but he didn't regret his decision. Being in civilization wasn't easy, but it was where he felt he needed to be. I'd finished the first draft of my fire tower memoir. Maybe it was time to start another chapter.

"Don't do it for longer than five years," the words of a lifer rang in my ears. "You don't want to lose your relevancy in the world."

Her words surprised me. She'd worked on towers for nearly twenty-five years and it made me sad to hear her say that.

What does it mean to have relevancy in the world? I recalled my earliest concerns about the fire tower. I'd been afraid of losing relevancy, becoming forgotten by people I loved, feeling unseen by the world. I'd been so afraid to slow down my pace of living. Retreating to an off-the-map cabin in the woods made me worry about losing my sense of visibility and community, of being someone whose work mattered. But over the years, I'd discovered I felt deeply relevant at the fire tower. I had a unique sense of purpose. I loved my job. I'd learned, the hard way, how to watch for smokes—"catch 'em small"—and call them in quickly. And while I was physically apart, I wasn't alone. I felt deeply held and supported by a community of lookouts, dispatchers, firefighters, and pilots. I belonged to something unique and much larger than myself.

My wildfire friends—lookouts and firefighters alike—had begun to joke that I'd become a lifer, but I always laughed off their words. I thought of the veteran lookout's advice and shook my head.

"I'll eventually transition back into 'real life,'" I'd say. "One day, anyway."

Hungry for a new challenge, I decided to opt out of a fourth summer of solitude and instead pursued a firefighting job—a more social experience. I trained all winter for the standard wildland firefighter fitness test, which involved a series of equipment hauls up and down a 45-degree-angle ramp, and was ecstatic to pass the first test in January. But when I had to take the same test again in March, I failed, my 115-pound frame collapsing under the weight of pulling an eighty-pound sled with a fire hose. I completed the test but finished thirty seconds slower than the time requirement. I was crushed. Now I was out of a firefighting job *and* a tower job.

Two weeks later, on the morning of my thirty-fourth birthday, I received an unexpected email from my former tower supervisor.

"We've just had a tower open up, last minute. It's yours if you want it."

My fingers hit Reply within sixty seconds of opening the email: "I'm in."

There was a catch: the fire tower they were offering me was a five-month season. I'd never worked longer than four months at the tower. I'd always said I could never do a five-month stint, the isolation would be too intense. But I wasn't ready to turn my back on the fire season and this strange professional subculture I'd stumbled into three years earlier. I'd tough it out, I told myself.

My new tower is positioned atop a steep hill in the clearing in a forest. This forest is different from the swampy wetlands and old coniferous forest of my former tower. Here, I am surrounded by a blend of young deciduous growth and old-growth trees. I can see nearly a hundred kilometres to the north from the ground, owing to recent logging of the trees on my north-facing slope. Hip-high stands of last summer's yarrow, cured brown, and wild raspberry and young willows spring from the clearing. The tower sits at a slightly lower elevation than did my previous tower, meaning it will probably be a few degrees warmer—which I don't mind in the least. It's considered one of the busiest towers in the district, with less rainfall, a higher tendency to drought, and several farming communities nearby, all of which increase the probability of wildfire ignition and spread.

Just take it one day at a time, I tell myself. I've learned that you can't rush the fire season. Not as a Lookout Observer, anyway. You have to take whatever happens—weather, wildlife, and wildfires—day by day.

And here now, on my first morning—eyes trained on the splendour of light sizzling in the morning sky—I can't help but wonder if, while I didn't choose the fire tower this season, maybe the fire tower chose me.

The sky on fire is a premonition for the months to come, my wildest season yet.

April is strangely meditative. I climb the tower every morning with a Thermos of coffee, books, my ukulele. The days are long, quiet, but I don't mind. I'm a sky monk now. Also, I know what's coming: May in the boreal, a lion's roar. The winds will inevitably whip through the stick-dry forests. Farmers will be out clearing their fields. There are so many ways that wildfires can ignite and let rip across the forest.

I want to enjoy the calm before the shit hits the fan.

Since it's early season, I climb down at four o'clock in the afternoon. I'm serenaded by Holly's joyous call: *AroooOOOOOOOoooOOOOOO!* Meaning, no doubt, *Let's go stretch our legs, Human!* Armed with a canister of bear spray, I follow a trail into the cutblock to the north, ready to discover the forest around me, where the game trails come and go, where the patches of wild berries spring from the earth. I spy tiny dots of electric green sprouting up from the bare grey soil—fireweed, I notice, taking root, reclaiming the exposed soils. Holly sniffs fresh wolf tracks pressed into the earth, tracks much bigger than my hand. Piles of hairy wolf scat are littered along the trail. I flash back to our surreal encounter with the black wolf two years ago: those yellow eyes boring into me.

The trails are wet, spongy, saturated by snowmelt.

My thoughts drift to Jay and my whole body misses him. After my second fire season we spent the fall and winter together. We drove four thousand kilometres east to a cabin in western Quebec, then five thousand kilometres back west to Prince Rupert to board a ferry for Haida Gwaii. On the north coast, on the edge of the Naikoon eco-logical reserve, we moved into a tiny cabin without running water or electricity. We cared for one another, but it was hard to live together. Maybe we were both too wild in our own ways. Maybe we were trying too hard to play house, to force ourselves into a mould that neither of us really wanted or were ready for. We aren't a couple anymore, but we're connected in a way that's difficult to name or make sense of.

A groundhog emerges from her den and furtively sniffs the air.

Humans are one of the few mammal species that choose to den together with their partners, their mates, for life. Bears mate in June and disperse into the bush. Mule deer does dwell mostly in small matri-archal groups, and come together with bucks during the late autumn rut. Many species of boreal owl pair up to breed for only several seasons before flying elsewhere. Even when they're sharing a nest, though, owls spend most of their days and nights alone, flying through the forest and meadows, hunting, procuring food for their young.

The groundhogs live on opposite sides of my hill. A big boar lives on the south-facing slope, while the sow has dug her tunnel on the north-facing slope. They live close but separately.

Part of me regrets that Jay and I couldn't make our relationship work. But on better days, today being one of them, I can accept that nature works in mysterious ways.

I've learned that it's okay to deviate from conventional pathways.

Deviance is a part of evolution too.

By late April, tender green shoots are rocketing up out of the soil. Fat dandelions proliferate on the southern slope. A bear emerges from her den, a startle of black in the northern cutblock. A large sow, probably three hundred pounds. She feasts on the young fireweed shoots and sticky willow buds, with two yearling cubs grazing alongside her. I freeze on the spot. I'm surprised to see them, but I shouldn't be. The former lookout kept meticulous records over ten years, and he'd recorded seeing twelve different bears at this tower in the span of a single day.

The bears remain at a safe distance from us—three hundred metres or so. Holly doesn't even seem to realize they're there. I check the electric bear fence that surrounds my yard—thankfully, it's working—and watch the sow lumber a few steps with her head pressed to the earth, jaw grinding down greens. Black bear cubs generally stay with their mother for eighteen months. They'll spend the early days of spring and summer together, feasting and foraging. But come midsummer, the sow will likely chase off the cubs to make their own ways in the world.

An hour later, I go outside to use the outhouse. My heart nearly stops as I look down the southern slope and see the sow—appearing much larger, much more dangerous up close—only a few metres beyond the fence. She's less than ten metres away from me.

"S-s-hit!" I stammer loudly, and she yanks her head up to look at me. Her eyes are small and black and beady. Neither of us moves. I back off slowly, to the cabin, my heart pounding.

"Hey Bear!" I yell.

She only looks at me.

Then I blast an air horn once, twice. HONK-HONK! Like a semi-truck driver blowing his horn just as you pass him. The loud, obnoxious sound doesn't belong out here. The bear obviously thinks so too, because she promptly swivels on her large hindquarters and gallops off into the bushes, cubs close behind her.

Over the next few days, the mother and cubs return to the southern slope, devouring the dandelions and fresh blades of grass. They always appear silently, emerging from the edge of the bush. I do everything in my power to deter her from getting close to the fence. I knock on the cabin window. Blast the air horn. Yell in my deepest, most intimidating sounding voice, probably sounding entirely unconvincing:

"HEY BEAR!"

Holly looks at me curiously, wondering what the fuss is about.

Up in the tower, I bang on the sides of the cupola and empty my lungs with screams:

"GET GOING!"

I don't want her coming close to the cabin. She seems interested only in grazing, but bears can be unpredictable. Furthermore, I don't want to stumble accidentally across her path, or come between her and the cubs. Interestingly, Holly does nothing. I'm not sure if it's because she can't see them from our vantage point up on the hill, or doesn't smell them, or is simply relying on *me*, the alpha, to chase them away. I'm just relieved she isn't chasing them, enacting the age-old story: dog chases bear, bear chases dog—right back to human.

I'd only had a handful of bear sightings at my former tower, and those bears fled at the slightest provocation; a gentle rap on the glass window, a barely uttered "Hey Bear," and away they bounded into the bush. But here, I am closer to the edges of grain and oat fields—food sources for bears—and I suspect some of them are human-habituated. Unafraid of people.

A week later, I'm standing ankle deep in a mound of dirt, dropping potato seeds into the churned earth. I look up and see her watching me from the edge of the bush. The garden patch is outside the fence, so there's no longer a 10,000-volt electric current running between us. I call Holly, who's nearby, and she comes running to my side. Hoisting the garden hoe high in the air, I holler, "GET GOING, MAMA!"

She drops her head lower, eyeing me. The cubs are behind her, not moving, probably afraid. I slowly walk towards the cabin, Holly glued to my side, watching me, waiting for my next move.

"PLEASE LEAVE!" I yell desperately.

And then, at my wits' end, I drop my pants and urinate, right there, about three hundred metres upslope from the sow, as if to say, *This is mine.* The bush belongs to her, but my life also depends on this clearing in the woods. Before the introduction of electric bear fences at fire towers, some lookouts used to urinate around the perimeter of the yard, swearing that the scent kept bears away. I know how ridiculous I must look: squatting on a hilltop, the strangest kind of face off with a black bear. She gets my message—but doesn't back off. Instead, the bear stomps the ground with her two front feet and lets out a loud sound: CHUFF! Then, head low, she walks towards me.

I yank up my pants and speed-walk into the cabin. I know what I need to do, but I don't want to do it. I never actually thought the moment would come when I needed to rely on the 12-gauge shotgun. With trembling hands, I load a rubber slug. I'm not going to shoot her, but I need to scare her.

She's come closer now, head down, grazing unperturbed. She clearly doesn't see me as a threat.

I take a deep breath and aim a safe distance to the east of her. I pull the trigger and the gun bucks. The cubs run for a huge white spruce and scale the tree. Their small claws sound like peeling Velcro as they dig into the bark and climb. Mama dashes back to the treeline, then spins and faces me.

CHUFF!

She's pissed.

"Fuuuuuuuuck."

She's not going anywhere until the cubs come down. One cub is nearly twelve metres up the tree.

I'm not sure what to do, so I call a fellow lookout, Justin, the one whose family used to own Holly. Justin is an avid naturalist, hunter, and woodsman, and his tower is also frequented by bears. He'll know what to do.

"Go back into the cabin and give her some space to let the cubs come down," he says. "She'll probably clear off when they climb down."

I'm frightened, so I keep him on the line, waiting, watching the cubs through the cabin window. A few minutes later they climb down. I lose sight of the sow but imagine they've now pushed off into the bush, and step outside to check. Peering over the edge of the hill, looking down, I realize I'm five metres away from the sow.

"Holy shit! She's actually come closer!" I whisper in a panicked voice.

"Okay, you've got a habituated bear on your hands," says Justin. "She thinks your yard is her territory. You've got to give her a stronger negative consequence."

"Ugh, I don't know if I can do this," I admit, even though I know I have no choice.

I palm a rubber slug into the shotgun chamber. My dad had given me these slugs before my first season, three years ago, and I'd never had to use one before. Biologists use rubber slugs to deter habituated wildlife. "Aim for the rump," I remember him telling me. It wouldn't fatally injure the bear, but it would send a clear signal: it's not safe for you here.

I feel like a jerk, standing there, gun ready to be fired, but in the moment I sense this is the right thing to do. Habituated bears run the risk of being shot by farmers or ranchers, who would rarely opt for rubber bullets.

I take a deep breath, aim for her large rump, and fire on the exhale.

Mama jumps and gallops downhill towards the bushes, the cubs hot on her heels. She doesn't stop. I hear them go crashing through the bush.

I don't feel triumphant, standing there with a shotgun in my hands. A part of me feels awful. Another part of me feels I've done a kindness to the bear. Truthfully, I'm not sure.

My heart slows and I find my breath again. I stare hard at the empty space they've left behind.

May is here, and it's been unusually hot in the North.

"Something is going to happen today. These clouds want to do something," I text Grant.

Grant is one of the firefighting crew leaders who's become a friend over the past couple of years. He and his helitack crew have been stationed by a nearby fuel cache, just a few kilometres northwest of my tower, on a five-minute getaway. They'll be patrolling the forest around my tower for added precaution, though my eyes are scanning constantly; I'm determined not to miss anything.

"We shall see," Grant says in his calm, cool-headed way.

For the past week and a half, we've been designated on high to extreme fire hazard. We're all waiting for the inevitable: a wildfire that wants to tear across the kindling forest. It's so windy that the cupola rattles back and forth like a maraca.

But by six o'clock in the evening, nothing has happened. Maybe Grant was right.

The phone rings. It's a good friend from Peace River. As we catch up, I watch the clouds churn and darken, their bottoms flattening out. Several clouds surround my tower. I've discovered that listening to a voice while up in the cupola actually makes me better at my job. It helps to focus and train my eye. Even though I've been up in the sky since nine o'clock this morning, my eyes feel razor-sharp. My friend is confiding to me about the uncertainty she's feeling about her job when out of nowhere a bolt of lightning splits a cloud and stabs the earth—less than a kilometre away. My eyes nearly pop out of my skull.

I can feel the snap of electricity. Thunder crashes like a grand piano down a flight of stairs.

"AHHHHHHHH!" I scream.

"I heard that! I felt like I was up there with you!" my friend cries.

My whole body trembling like a live wire, I hang up to call in my first strike as another dry strike spears the earth. And another. This is an active storm, one that I know is going to cause a smoke.

"Did you see that? Dry lightning!" Grant texts. They can see the storm from the ground at the fuel cache to the north.

"Yes! Calling it in right now."

Grant's crew is dispatched to investigate the storm, hunting for low, hidden smokes. But they can't get close enough—it's far too volatile, too many lightning strikes—and so the crew is dispatched back to base for the evening.

I stay up past 8 p.m., well beyond the call of duty, to keep vigilant for smokes. The storm pushes to my southwest towards the town of Manning. I watch the dry strikes continue, but it's pointless to call them in. The duty officer knows: this is an active storm. We're going to get wildfires, if not tonight, likely tomorrow during the peak heat of midday.

Justin calls. He's still up in his fire tower too, watching like a hawk.

"I don't think I've ever seen lightning like this so early in the season," I say, pacing the cupola like a caged tiger, charged with adrenalin.

At 9 p.m., nothing has happened—no smokes. Justin and I sign off for the night, agreeing that we're better off climbing down and getting some sleep.

"Tomorrow is going to be a crazy day," he says.

I hang up, pull the worn red harness over my shoulders, and gather up my belongings to climb down for the night. I look up and see the smoke: a single black column, rising.

"Wow," I say to nobody.

It's late in the evening for this kind of ignition. The smoke is tar black—must be burning in pine and spruce, a highly volatile fuel type. Hands trembling, I spin the Fire Finder around and line up the shot. I don't take the time to estimate the distance or fill out the pre-smoke form; I simply pick up the mic and call in my bearing. The smoke is rising over a ridge and I'm uncertain of the distance. Could be thirty, forty kilometres? It doesn't matter. They'll see it.

"Pre-smoke," I say into the radio mic. The smoke intensifies, rolls and braids itself into a heavy black undulating rope. This is bad, I think. This is going to be a fuck-off fire.

Justin calls back within a few minutes. "Damn," he says, breathless. "You just beat me to it."

Turns out we had both called in the same smoke—even though there's more than 150 kilometres between us. I take his cross-bearing and map it out. My distance was way off. The wildfire is burning more than seventy kilometres away from me.

"It must be the first time in history that our towers have crossed off on a wildfire!" he exclaims with a laugh, and it's probably true. We don't know it yet, but we'll be watching this fire burn for the next two months.

Suddenly, the lookout who used to man my tower jumps on the radio. He's heard our late night detections and hurried up his tower. He's closer to the blaze than Justin and I are.

"There's a second smoke!" he says.

And sure enough, I spy a second column popping up over the ridge.

The text messages from the firefighting base in Manning start pouring in.

Eric sends me photographs of the wildfire, burning less than twenty kilometres north of the base. He's standing on the porch and the charcoal smoke rises up behind him, staining the sky. He's just returned from his days off, and since he isn't on duty, he wasn't dispatched to the fire.

"Every single firefighter on base has just been dispatched to the smoke!" he writes.

Later, one of the firefighters will tell me it was the wildest thing he'd ever witnessed in his entire firefighting career. The largest smoke column turned bright amber in the setting light. The wildfire was burning deep in the bush and there wasn't any road access to get close to it. It was, indeed, a fuck-off fire, too hot and volatile, far too dangerous, to get closer to. Crews would have to wait until morning, until there was enough light to safely fly helicopters and to get air tankers fighting the fire from above, dropping loads of retardant on it.

I stay up until 11 p.m. watching the smoke column become a monster on the horizon, fear rising in my chest. The lookout closest to the fires, only a few kilometres away, is gathering her belongings and about to evacuate the tower. She sends me a quick text to notify me. Fortunately, a dirt road connects her to civilization, so she can drive into the Manning fire base. She's a lifer, though she's never been evacuated before.

"That must've been so scary," I say to her the following evening, after she's been evacuated.

"Oh, it wasn't so bad," she replies quietly, unfazed—the epitome of calm. "The frogs were singing and the moon was beautiful. I wanted to climb the tower to get a photo of the fire, but there wasn't time."

I climb down before midnight and will myself to sleep.

Tomorrow will be another day of burning.

The next day, gusting winds push Fire 52 to more than five thousand hectares in size. Multiple tanker groups are dropping loads of retardant to prevent the spread into Manning and surrounding rural communities. Helicopter pilots drop buckets of water on the hot spots, dozers clear wide tracts of bush to create breaks, and firefighters start back burns in nearby fields to prevent expansion.

The lookouts watch from afar, wide-eyed, mesmerized by the smoke. We're all convinced that something else is going to ignite today, but as it turns out, it doesn't happen in our district.

By midday, I spy a faint grey smoke coming up far to my north. I guess that it's well over a hundred kilometres away, across the district line and into High Level, but even so, I line up the shot and jump on Channel 99. I call up my northerly neighbour and pass him my bearing.

"Thanks," answers a deep male voice. Sean has been a lookout for nearly twenty years. "We called it in about ten minutes ago and crews are already on the way."

The smoke is barely visible from my tower, just a slight mark on the horizon, as though someone has taken an eraser to a dark pencil marking. In my rookie season I wouldn't have had the eyes and experience to be able to distinguish it as a wildfire.

I don't know it yet, but this eraser smudge on the ledger of the horizon will become the Chuckegg Creek Fire and grow to an unfathomable size—triple the size of Calgary.

Bigger than Luxembourg.

Bigger than Hong Kong.

———

We stay on extreme hazard for three weeks. Every day starts at 9 a.m. and ends at 8 p.m. I can't imagine how I would've fared had this been my rookie season. Finally I understand the reason they asked me in my interview, three years ago, "What do you do to treat eye fatigue?"

I can't focus on a book, or write, so I listen to music and do laps in the cupola like a frantic squirrel. I talk to other lookouts on the phone and trade information on Fire 52, which they're now calling the Battle River Fire. Forestry has designated it a "complex fire" and has set up a temporary fire camp north of Manning. Hundreds of managers, firefighters, and wildfire personnel have travelled from other parts of Alberta, and even from B.C., to help manage it.

My tower neighbours and I are feeling enormous pressure to stay vigilant on these eleven-hour days. The importance of being a lookout has never felt so real to me, the stakes never so high. The forest is a tinderbox, and the high heat and high winds are relentless, day after day. Weather reports continue to defer any expectation of rain in the four-day forecast. No rain in sight. Forestry issues a fire ban in the Peace Country, halting the burning activities of farmers, land users, campers, and recreationalists.

Only a week after the Battle River Fire and the Chuckegg Creek Fire ignite, gusting winds—rushing at over fifty kilometres an hour—intensify the burning conditions. The temperature climbs into the high twenties, bizarre weather for May in northern Alberta, and I roast up in the Plexiglas cupola. By midday, when the temperature reaches 25 degrees Celsius, I watch, stupefied, as both wildfires take off. To my southwest, the Battle River Fire doubles in size—now ten thousand hectares. To my north, the initial eraser smudge has now become legible text on the horizon:

O-U-T-O-F-C-O-N-T-R-O-L

By dusk, both smoke columns are colossal. The one to the north is—by far—the largest wildfire I've ever witnessed up close; I estimate it's blown up to more than twenty thousand hectares. I photograph the massive column, lying flat on the northern horizon like a blue sleeping giant. It's a jaw-dropping sight, terrifying, and yet stunning against the dimming light. I'm inspired by this act of Mother Nature, but also deeply anxious regarding what's about to happen. The fire is burning less than twenty kilometres away from the town of High Level. I almost feel guilty taking the photographs, marvelling in this uncontested beauty, because I worry, if these high winds keep up, Forestry won't be able to do much to prevent the fire from reaching communities.

Eric's crew is stationed on five-minute getaway at my fire tower. They arrive, day after day, and camp out on my front lawn. By day, I watch the "megafires"—a term now commonly used by wildfire scientists— burning out of control and scour the landscape for more starts. By night, I climb down, hopped up on adrenalin, and mix the ingredients for chocolate chip cookies and wild berry scones for the crew.

"Trina, you don't need to do this for us," Eric says with a laugh. "We have plenty of food at base." I share with the guys plates of baked goods and pour mugs of hot coffee during my fifteen-minute lunch break on the ground.

"Oh, I know," I say. "It calms my nerves, though."

I love their company and moral support, though I barely get to interact face to face with them. Somehow, knowing that they're below allays my anxiety. I don't want to be alone, sandwiched between two megafires, tasked with the job of early wildfire detection during the craziest burning conditions I've ever experienced. We've now gone more than four weeks without rain.

The temperature hits 30 degrees by the third week of May. There's no shade from the sun on the hilltop, so I dip a leopard-print sarong in cold water and drape it over Holly's overheated body. She pants loudly, trying to cool herself down, and licks my hand appreciatively. "My poor little leopard," I croon to her. Eric and the guys bunch together in the small shards of shade cast by the shed. They're all shoeless, shirtless, overheating in their green woollen, fire-retardant Nomex pants.

Back up in the cupola, looking southwest, I spy a column of grey smoke coming up over the ridgeline, about twenty kilometres out from the tower.

"I've got a smoke, guys!" I holler down.

They race towards the smoke in the helicopter, though it's quickly called off.

"False alarm," says Eric over the radio.

I'd accidentally called in a hot spot burning near the Battle River Fire. It's a tricky game to identify smokes near the megafires, particularly from a fixed point in the distance, How to know whether it's part of the complex fire or a new fire? I'm unfazed by my mistakes. They're just mistakes, nothing more. Better safe than sorry, as the Happy Lookout once said.

On May 26, the Chuckegg Creek Fire, the one to my north, burns within five kilometres of the town of High Level. Photographs of the massive smoke columns spread on social media. The duty officer from High Level calls me every hour to ask for a wind report. Which way is the wind blowing? How fast? Today the winds are gusting at sixty kilometres per hour from the south—bad news for residents of High Level, as the wildfire is threatening to push even farther north and raze the town. Everyone is bracing themselves for the worst, praying for a sudden wind shift that will push the fire off its trajectory.

The smoke column now looks as though a hydrogen bomb has gone off. It has risen to a staggering atmospheric height, forming a cloud that meteorologists call pyrocumulus. Cumulonimbus flammagenitus, or pyrocumulus, or a "pyroCB," forms above a heat source—such as a wildfire, or even a volcanic eruption—and can actually produce its own weather system, including vicious winds and lightning. Such an astonishing event happened with the Horse River Fire in Fort McMurray in

May 2016. A pyroCB formed over the blaze and produced lightning that ignited several new fires over forty kilometres northeast of the fire's front.

Residents of the farming communities to my north—only fifteen, twenty-some kilometres away—are ordered to evacuate their homes. The order is mostly precautionary, due to the thick smoke smothering the horizon. The smoke reaches my tower too, depending on the direction of the wind. It's thicker than I've ever experienced before, leaving me feeling vulnerable, unable to locate any new wildfires. I commiserate with Eric's crew, who are forced to leave their man-up post early because of the thick smoke choking the sky and making it dangerous to fly.

Three days later, the unexpected happens.

The winds shift from the southwest to the north, pushing the colossal fires south. By noon, the 100,000-hectare fire is making a mad dash towards the recently evacuated farming communities north of me. I watch, horrified, as the smoke rises even higher into the sky.

Sean, the lookout to my north, decides to self-evacuate.

"Hey, I'm gonna hit the road. The fire is about five kilometres away now," I hear him tell his neighbour over Channel 99. "And I'm about needing a gas mask at this point."

Stupefied, I watch the fire double in size—four massive smoke columns rising to staggering heights. It now stretches across my entire northern quadrant.

Then I glance down and see her, the black bear sow and her two cubs, grazing comfortably on the southern slope, just metres away from my cabin.

Impeccable *fucking* timing.

"GET GOING, MAMA!!!" I holler down at them. I bang the cupola walls like a madwoman, then glance back up at the pyrocumulus that's morphed into a monstrous, black-bottomed thunderhead. It seems as though it's on the brink of producing lightning.

The bears don't budge, unbothered by the irate lookout, unbothered by the possibility of the forest burning down. The bears do what they always do: fill their huge maws with food. The sow flops on her belly and feeds lying down, as if to say to me: You want me to move? Make me, lady.

I make a split-second decision.

"XMA26, this is XMA685," I say into the radio mic. "Can you mark me away from the radio for five minutes? I've got to deal with a bear below."

"Copy that."

Wearing nothing but a pair of jean shorts and a sports bra and my hiking boots, I throw on my harness and sprint down the ladder. I rush into the cabin, load a rubber slug in the shotgun chamber, and find a good position on the sow. No time to be afraid, not when a 200,000-hectare fire is coming right for you. I line up the shot on her rump.

Fire.

The sow jumps and makes a beeline for the bush. Her cubs follow suit.

"GET GOING!" I yell in my deepest baritone voice.

Then I laugh hysterically, because this day could not get any more intensely bizarre. Forget Rapunzel, or Tower Girl—just call me Goldilocks with a Gun.

By dusk, I can see orange flames pulsing at the base of the Chuckegg Fire. The gusting winds push the fire south. It's now forty kilometres away from my tower. I text the duty officer photos of the colossal smoke columns, at least thirty of them. Every few minutes a new puff of smoke at the base threads itself into the larger column. I can hardly believe that this view of a disaster in the making is mine and mine alone. I look into the farming communities to the north, wondering whether everyone has been evacuated.

The duty officer calls me from the office in Peace River. Although nearby communities are being evacuated, it's mostly owing to the smoke hazard. The Chuckegg Creek Fire remains a long shot from my tower. Most lookouts won't be evacuated until a wildfire comes within ten kilometres of their tower, sometimes even closer, although smoke hazard poses another concern. But this is a 200,000-hectare megafire spreading at an alarming rate. The duty officer tells me to be ready for a worst-case scenario.

"Better pack your things tonight—just in case we need to get you out of there," he cautions me.

I climb down and pack my "bug-out bag," an emergency bag with my laptop and camera, ID and wallet, some clothing and toiletries, and dog food for Holly. Not knowing what else to do, I sweep the cabin floor and water my tomato and basil seedlings. Then I open cans of peaches and decide to bake a peach cobbler.

A fucking peach cobbler.

Who am I? The Martha Stewart of fire tower lookouts?

I know it's absurd, but the task distracts my anxious thoughts, busies my hands. Through the clearing to the north I can see nearly the entirety of the Chuckegg Fire from my kitchen window. I mix flour and sugar and butter and cinnamon and layer it in a baking pan over sliced peaches—glancing up every few seconds to take in the raging inferno moving closer.

I wake early in the morning to the smell of smoke and rush outside. The world glows an eerie greyish yellow, a post-apocalyptic shade. The air quality is abysmal; I know I'm breathing in all kinds of toxicity. I can't see beyond the stand of trees bordering my yard. Ash cascades from the sky into my hair and onto my shoulders. I hold out my hand and collect the falling paper moths. My first thought: How close did the wildfire get last night? I retreat to the cabin and try to go back to sleep, but I'm afraid of what I don't know.

A friend texts me. "Are they evacuating you, or what?"

"I'm not sure. The duty officer will call me soon."

"If you hear the fire's roar, you've gotta run."

Justin calls. "Worst-case scenario, you can run to the beaver pond," he says, referring to the pond about a kilometre south of my tower. "Swim out into the middle of the pond."

I try to stay calm. No news from the duty officer is good news. I trust the system. They know I'm out here and there's no need to panic. I take a few smoke-filled, deep breaths and wait for the call.

Eric calls. "Don't worry, Trina," he says, his voice steady and calm, even cheery. He and the firefighting crews just got off their early morning conference call with managers. He reassures me there's a plan in place. "No one will get left behind," he insists.

The phone rings again.

"Trina," the duty officer says over the phone. "We can't believe it, but the Chuckegg took a twenty-two-kilometre run south last night."

The wildfire did what no fire scientist had anticipated: burned over fifty thousand hectares in the span of a single night. Wildfires typically calm down at night, slowed by lower temperatures and higher humidity. It even jumped the kilometre-wide Peace River, throwing burning embers across the brown, snaking waterway. At the same time, the Chuckegg Fire burned over more than nineteen homes and properties in one of the northern farming communities. Fortunately, people were already evacuated, but tragically, many families lost pets and livestock left behind.

"Listen, don't worry out there," says the duty officer, sensing my rising fear. "We've got a trigger system in place. But the fire is still twenty kilometres away from you. That's a long way off. If I were you," he adds gently, "I'd be more worried about the bears than the fire."

I summon up a laugh. He is probably right.

I stay on the ground. There's no point in climbing back up to the sky. For Holly, it's just another day at the tower. She follows me like a shadow. We retreat from the smoke into the cabin.

I brew coffee, clean the cabin, and pack up my belongings—just in case.

By noon, managers in Peace River make the call. The duty officer has given the signal to evacuate me from my tower. One of the rangers will drive in with a 4x4 on the makeshift road and pick me up. The reason for the evacuation is the intensity of the smoke blowing in from the Chuckegg, choking me and the sky, rather than the threat of wildfire.

I am relieved to see him pull up a few hours later. He loads my gear in the back of the pickup and seems relaxed about the whole evacuation. It isn't a big thing, he says. "The Chuckegg is a long ways off." He's more frustrated that they have to dismantle the camp, because he'll likely be ordered to build it back up in a few days' time. He would've rather stayed and endured the conditions. If there's anything I've learned, it's that firefighters don't mind the smoke. To them it means action, work, and the opportunity to do what they've been trained to do.

Holly jumps eagerly into the back seat, pink tongue dangling. She's always game for a joyride. I run back into the cabin to grab my potted coffee plant.

The highway is choked with smoke from both the Battle River Fire and the Chuckegg Creek Fire. It feels like a ghost highway, barricaded off in Manning and farther north around High Level. We drive for twenty kilometres without meeting another vehicle, the smoke so thick we can see only ten metres in front of us. We pass a cinnamon-coloured black bear, grazing calmly in the ditch, and it occurs to me that it feels more normal to see a bear than the reflective road signs we pass, which—in my imagination—read WELCOME BACK TO CIVILIZATION, TRINA.

The haze follows us 150 kilometres south to the town of Peace River, completely obscuring the river valley and the rolling hills.

I had hoped he'd take me to the Manning fire base, where Eric and Grant and my other firefighter friends are stationed. At least there would be camaraderie there. But the ranger drops me off at the front door of my parents' empty house in Peace River. They are away visiting my brother's family in Edmonton. I blink, unbelieving, in their front entrance, the hardwood floor gleaming. The house is clean, polished, sanitized—dust and dirt free.

I feel like a bear that's been tranquilized and relocated to an unfamiliar territory. As the adrenalin of the sudden evacuation wears off, I wake up, realizing that I'm back three months earlier than anticipated. I don't know what to do with myself. I'm "bushed," as my dad would say. Culture shocked. Stripped of my purpose of watching for smoke.

I take a hot shower—with *real* water pressure—and the dirt and ash that's accumulated on my skin and hair dissolves at my feet and slides down the drain. I crawl under the plush duvet covers in the guest room, Holly curled at my feet, and fall fast into a deep hibernation.

The very next morning, I drive to the Forestry office in downtown Peace River and ask the duty officer: "When can I go back?"

He points to the image on a screen, an image taken by the digital camera that's positioned atop my fire tower. It's as if someone has wrapped the camera in thick grey wool. Black specs, probably ash, cling to the lens. He doesn't have an answer for me.

I drive up to the supply warehouse to pitch in. I wash and bleach over fifty water buckets. We'll fill these up and seal the lids with a hammer. These are the buckets that keep lookouts alive during the long dry spells in the fire season. It's monotonous, humbling work. I'm relieved to pitch in behind the scenes and help with these unglamorous tasks.

The work distracts me from my new sense of worry. How long am I going to be away from my fire tower? The truth is, I *want* to go back. Desperately. I think of Justin and my other colleagues out there, witnessing the megafires, keeping vigilant for new ignitions. I even think of the sow with her cubs.

The modern comforts of civilization—hot water, electricity, access to grocery stores, restaurants, movie theatres, and shopping malls—aren't tempting. Not when there's a world of adventure—the proximity to nature and the deep sense of purpose I feel at the fire tower—beckoning to me.

My mind travels back to a conversation I'd shared with Dan, the new lookout at my previous fly-in tower. Dan seemed perfectly suited for lookout life: a musician and wildlife photographer who spends hours wandering the bush, lying flat on his belly, following songbirds with his lens. On his first morning at the tower he captured a stunning photo of a lynx prowling the edge of the airstrip. Maybe an omen of sorts. Dan seemed to belong to lookout life without even trying.

One afternoon, before the megafires took off, Dan and I were talking on the phone, tower to tower. I heard myself lamenting to him about how far I'd strayed from the life path I'd envisioned as a child, the one I carried into my twenties. I thought I'd be a doctor, or teacher, a human rights activist, a wife, a mother. When in fact I'd stepped into the story that had once seemed more like a fantasy.

I'd become a writer and a lookout. A woman who chose to be alone in the woods.

Deep down, I did wonder, Am I a lifer? What had I made of my life? On days when I doubted myself and my choices, I worried about my path. It appeared that I'd become a crazy dog lady who watched for smoke, peed in a yogurt container and flung it out the window, picked wildflowers, and yelled at bears using commands of "please" and "thank you"!

"Maybe it's time to go back to the real world," I lamented.

"Trina, this is the real world," Dan said, cutting me off. "The world out there isn't always real." His usually quiet voice became fiery. "There's a lot of fakery and hiding. Sometimes it feels like it's made of Saran Wrap."

I laughed. *Saran Wrap!*

"But this nature that surrounds us, it's real, Trina."

I nodded my head, looking out on my expansive domain of forest, watching a northern harrier circling then diving and gliding above the treetops, hunting for songbirds.

Over the past couple of weeks, Peace River has become a refuge for evacuees. Thousands of people have left their homes and properties in High Level, hundreds have fled the rural communities surrounding Manning.

My vantage point has changed from looking out on forest and wilderness to looking into the lives of those people affected by the wildfires.

While getting my sun-fried split ends cleaned up at a salon in town, I hear stories of generosity and camaraderie. People in Peace River have opened up their homes and businesses to support the evacuees. Residents offer them free haircuts and meals. One trucker donates his services to help relocate entire herds of cattle from out of harm's way.

People band together during a crisis. Maybe we realize what's important. There's less hiding, more vulnerability, and we become less sealed off from our shared humanity. I also overhear people, everyday people who are not at all connected to the wildfire community, talking about what they're seeing and how they're affected by it.

"Is this the new normal?" a woman asks her friend in the supermarket parking lot. "Are we safe here? Will the wildfire reach Peace River?"

Smoke dominates the skies, the landscape, people's imaginations.

Ecological grief. Anticipatory grief. Climate grief. This kind of grief smoulders beneath everyday comments about the wildfires. We are trying to make sense of forces that feel completely out of our hands and beyond our collective control.

Their words follow me, only a few days later, as a ranger drives Holly and me back out to the fire tower. We pass through the police barricades and drive north along the ghost highway. Evacuees won't be allowed back to their homes for another week or so. Everyone is praying for rain to quell the spread of the fires, to cleanse the smoke particulate from the atmosphere.

The truck ambles along the rutted dirt road that leads to my tower. A large black bear scoots across the road and I realize that I'm happy to see him. The forest is burning, but the bear doesn't mind. New growth will rise from the ashes. Food for the bear. Food for many ungulate species in the forest. There isn't a season for worrying in the bear's world, only hunger and the need to feed. The bears adapt, as humans will be forced to adapt, to an inevitably changing environment. How we choose to adapt remains to be seen.

I've been away only a few days, but somehow the fire tower appears wilder, as though half swallowed up by the forest. The dandelions have gone to seed in the tall, uncut grass. My potted tomato seedlings have grown fat and leafy in the persistent heat. Everything in the cabin is as I left it. The peach cobbler, preserved in the freezer.

I'm relieved to be back home where I belong.

Within an hour of arriving, I hear an enormous clap of thunder, so I hurry up the ladder to the cupola, eager to get back to my lightning-and-wildfire-watching duties. The texts from my lookout neighbours start pouring in. CHIME-CHIME-CHIME. Sounding the bells of my return.

"Glad to have you back," my friends tell me.

The Chuckegg Creek Fire burns all summer along, reaching the size of 350,000 hectares. The winds push the fire this way and that. One day it takes a giant run east towards the Mennonite community of La Crete and people are once again forced out of their homes. Impressively, Forestry is able to prevent major infrastructure losses by creating fire-guards around communities. I'm awed by the efforts of the Alberta wildfire community to manage the megafire and keep people out of harm's way. Hundreds of firefighters are imported from B.C. and other Canadian provinces. Some come from as far away as Mexico and South Africa to help quell the spread of the Chuckegg Creek Fire. There are so many individuals working hard on the front lines of the fire, with even more people hustling behind the scenes—and I'm inspired by their work.

It almost becomes normal to look out my window and see smoke rising to the north. Almost, but not quite. I joke that I should probably cover my northern windows with black garbage bags so I no longer see smoke and worry, Is it part of the Chuckegg or a new smoke? Inevitably, being only twenty kilometres away from a complex fire, where dozens of fires are burning every day, I call in a handful of smokes that turn out to be part of the Chuckegg. Some days I feel helpless, watching the monstrous smoke, knowing there's nothing I can do but observe. But it's also a humbling feeling, recognizing that simply playing witness is enough. It's my job.

And one day in late June, I do locate a new wildfire—one caused by lightning—which I estimate is just outside the Peace River district, in the High Level district. It's the tiniest puff of white smoke along the eastern edge of the river. Like a fleck on a fingernail. Unmoving. I watch it, uncertain. I haven't seen smoke behaviour quite like this before. I decide to call it in directly to the High Level district.

My tower neighbour to the northeast can't see anything. That doesn't bode well for my detection, since she's a lifer and a keen spotter. She doesn't miss anything.

I listen over the radio as the crew flies towards the smoke.

"Confirmed wildfire," I hear the leader's voice say. I listen for the exact GPS location of the fire and realize I'm less than one kilometre off. This wildfire is thirty-one kilometres away from me. I feel a swell of pride.

"It's burning in aspen," he says over the radio, which explains why the smoke appeared so white and still. Wildfires don't typically ignite in aspen, which is a green tree, its leaves full of moisture. It's indicative of just how dry the forest is. To my north, the drought codes, a numeric rating of the average moisture content of deep, compact organic layers, are sky-high and setting records. The code is an indicator of seasonal drought effects on the forest, and of the extent to which fire can burn and smoulder deep beneath the earth. The lookouts in High Level have been on extreme hazard alert for nearly the entire summer.

We're dying for the June monsoon that never materializes from the unforgiving skies.

———

The rain doesn't reach the North until late July. But when it comes, it comes hard, slapping the earth, flooding the fields and ditches. The season for new ignition is drowned out, just like that.

Big drought, big fires, big rains. We swing from extreme to extreme: fire to flood.

Eric texts me a video of the crew's favourite activity these days, when they're off-duty in the evenings: ditch surfing. There's so much runoff water in the ditches, they're overflowing. I watch a video of Eric, dressed in his wetsuit, with his surfboard in the ditch. He's holding on to a slack line that runs to his crew mates in a vehicle. As the driver accelerates, Eric stands up on the board, cruising the ditch. I hear cheers and laughter, and I'm laughing too, wishing I could witness these shenanigans in person.

"The fire season is effectively over," jokes one of his crew mates. "But our ditch surfing game is strong."

Despite the megafires burning to my south and north, it's been a relatively quiet fire season for me. I thought I'd have a record year in terms of number of smokes reported, but it's been the quietest season since my rookie year. I'm slightly jealous I don't have a ditch to surf.

Instead, it's been the season of bears. Bear season. Never in my life have I seen so many black bears. I accept that I'm living in a natural wildlife corridor; the bears have probably been coming here for hundreds, maybe thousands of years. The mother with yearling cubs appears frequently, although she seems to keep a wider distance between the bush and the fence. My tolerance has increased too. I'm growing more comfortable—cautiously comfortable—with the bears in my vicinity. I learn to set boundaries. Once they set foot up my slope, I try to deter them from coming closer: yelling, air horn, or firing the gun. In June, during mating season, I enjoy watching the mating rituals of boars and sows. From the tower, I see a huge boar saunter after a smaller sow for nearly two weeks. The sows go into estrus for only forty-eight hours, so the boars have to pursue them until they're ready to mate. Some days I put on a tune by the California Honeydrops for them as they skirt the edges of my life, giving me a front-row seat to their mating rituals: "All Day, All Night."

It's been my most social fire season yet. Through the latter half of the summer, friends and family members hike or drive in on ATVs, navigating the rutted-out, muddy road. They pitch their tents beneath the tower, gather around the campfire at night, and strum their guitars and ukuleles, marvelling at the view of the forest from my grassy hill-top. I feel at home here, hosting people I love and care about, and offering them a window into my life at the fire tower—a lifestyle I've slowly grown to love and cherish.

In late July, my parents come out with my six-year-old nephew, Masen, and my now two-year-old niece, Brielle. Masen scurries around the yard, following Holly on her scent hunt for the resident ground-hogs. They stumble upon a groundhog den and he hollers up to us, astonished, "We found their home!" We go walking along the road, lined with five-foot-high fireweed, and pick the blossoms for drying. My mother is enchanted with the variety of wildflowers: purple lung-wort, yarrow, yellow arnica, and the proliferation of big-headed sweet williams, broadcast on the slopes by a former lookout. My dad teaches Masen how to harvest old man's beard, lichen hanging from the low spruce boughs, and light a campfire—just as he taught my brother and me when we were kids. Brielle totters around the yard with a big grin, watermelon juice dripping from her cheeks.

Masen gazes up at the fire tower as I climb, eyes wide and wonder-struck. He takes enormous pride in tugging on the pulley-rope system and slowly sending my backpack up to the cupola.

"Auntie, maybe one day I can climb the tower!" he says earnestly.

"Maybe," I say a bit wistfully to my six-year-old nephew.

It's a gift, I realize, to be an auntie. To share my unique profession and home in the woods with my niece and nephew, to show them that Nature isn't a place to be feared, rather one to be loved and respected. And to know that there are so many different ways to live.

These days, I'm growing ever softer in my solitude, and making peace with the events and choices of my past. Some days I feel lonely, but even the loneliness is sweet—filled with a kind of nectar of meaning to tap into, a longing to make art, to write, to express myself. Slowly, I am learning that I'm strong enough, I am *enough*—alone, standing atop a hundred-foot fire tower, standing sturdily on my own two feet, despite the uncertainty of life.

I am surrounded by a beating pulse of visible and invisible relationships, a thriving ecosystem, a community of lonely souls longing for connection. And we are connected.

In Nature, I've come to understand, we are never alone.

By late August, the fire season is waning.

I return from a long run along the trail, sprinting the last several hundred metres back to the cabin, my pace quickened by trees bending in the sudden winds, branches snapping.

The clouds churning darker in the low-hanging sky. I'm exhausted from the run, but I can sense that lightning is imminent. I don't want to miss the show—probably the last of the fire season. I throw my harness over my shoulders and rush up the ladder to the sky. Holly curls up in her usual spot below the tower. The clouds are low, bluish-black, and ready to shoot strikes. I wrap my blue poncho around my shoulders and hit Play on my music.

Bob Dylan's "All Along the Watchtower" fills the cupola and I start to groove.

I don't need to be up here. Nothing is going to burn—it's far too wet in the forest.

I remember the words of Jack, the lifer I met before my first season, who told me about climbing the tower late at night to marvel at the Milky Way and a smattering of stars.

"I don't have to report shooting stars," he'd said wistfully. "They're just for me."

Perhaps my friends are right about me. Maybe I'm not so different from Jack, or Ralph, or the men and women who migrate back to the fire tower every season.

Maybe I am a lifer.

Who can really say?

Lightning splits the sky like an axe, and my eyes are wide open, watching.

ACKNOWLEDGEMENTS

August 29, 2020
Present weather: Romeo Whiskey
Days at the fire tower: 130
Cups of coffee brewed today: 3

Summer leaning into autumn and the cool days and star-scattered nights, the fireweed gone to seed and the black bears feasting on berries, make me already nostalgic for another fire season that's not yet gone.

It's difficult to acknowledge all of the individuals in both my literary and wildfire communities who have supported me in various ways over the past five years—you are many! What's certain is that neither of my childhood career ambitions—writer nor lookout—turned out to be very solitary professions after all. It takes a community to survive a season at the lookout, just as it takes a community to write a book. You remind me daily that "no woman is a fire tower."

Heartfelt gratitude to my publishing team at Penguin Random House Canada, and my editor, Amanda Betts, whose calm voice first found me in the thick of my wildest fire season. Amanda, thank you for facilitating such a gentle and transformative editorial process, and for grasping the heart of *Lookout* right from the get-go. I'm proud of what we've created together.

To my brilliant agent, Marilyn Biderman, and her sidekick Ruby, I am grateful for everything: the mentorship, tough love, shared excitement, and steadfast belief in my writing. It's been six amazing years together. Thank you for everything you do, Marilyn.

In October 2018, I attended the Emerging Writers' Residency at the Banff Centre for the Arts and benefited tremendously from the mentorship of author Kyo Maclear and the inspiring writers who made up the "Bower Birds." To Kyo and the Bower Birds, thank you for teaching me about the power of "Ma" and helping me to summon the courage to write memoir. I thought I was writing a book about wildfire and climate change, but as it turned out, it wanted to be a love story.

Special thanks to the Gushul Studio in the Crowsnest Pass, where I spent a month in a one-hundred-year-old house, metres away from the train tracks, editing the manuscript—and to Jenna Butler, fellow boreal writer, who joined me there. In February 2020, I returned to the Banff Centre for the Arts through the Writers' Guild of Alberta residency programme to finish the pages of this book. I also want to acknowledge the Alberta Foundation for the Arts for their financial support to work on the first draft of *Lookout*.

I'm grateful to a community of early readers whose eyes and attention to various details have helped to shape and fine-tune this book. In no particular order, thanks to: Melissa, Erin, Justin, Shane, Eric, Grant, Tori, Christa, Tiff, Jay, Marc, Ravi, Jessica, Harold, Mark, Dan, Meghan, Ashton, Nancy, Brett, Brianne, and Erinne. Thanks to Julia, Miles, Mitch, Kayla, Matt, Don, and Tara—who else am I missing?!—who tolerated my very many questions related to the history of lookouts, wildland firefighting, and helicopters.

To my dear, dear friends scattered in communities around the world, I'm grateful for your unconditional love and support. Thanks for letting me occupy your guest room, park a trailer on your property, or housesit, and for helping me to find ways to patch together an existence to write and rewrite drafts of this book. Mostly, thanks for sharing in the highs and lows of life together with hope, humour, and resiliency. I'm grateful for my adventurous pals who joined me at the fire tower over the seasons. Those treks in have meant the world to me.

To Eli, so many wild memories together: elk song, love letters, cross-Canada-150-road trips, and all. I'm grateful our paths crossed and for the many lessons our friendship taught me.

To my parents, my brother and sister-in-law, niece and nephew, words are not enough. Thank you for your unconditional love and support in a million different ways. I love you.

Atayo, *Alemisaru*. I wouldn't be who I am today without you.

To my dear cousin, Oscar. What are the chances that we grow up, ten years apart, on opposite sides of the country, and reunite at a fire tower in the northern boreal? Thanks for sharing in the adventures of wildfire—and life—together. You're more brother than cousin.

To my Alberta wildfire family: lookouts, detection aides, radio dispatchers, rangers, firefighters, pilots, and managers. You're far too many to name in these pages, but I'm grateful for the phone calls and radio calls, the flybys and man-up days, the rare cups of coffee, the shared excitement of cross-shots and catching smokes, and the unique camaraderie that's felt even from a lonely tower in the bush. I'm proud to work alongside you and play a tiny role in one of the best wildfire response systems in the world. You're the best part of every fire season. Special shout-outs to my supervisors, Mike, Jorge, members of HAC1 and HAC2 (resident man-up crews) and Eric "the Sky Pirate" for always looking out for Holly and me.

For all of those on Channel 99, if you're out there and still listening, I want to acknowledge the "Grandfather of the Forest" Ralph Cowie and his lovely wife, Bea, Don, Dianne, Sheila, Lila, Paul, Di, Joanne, Ruth, and Sharon—the many dedicated lifers who've helped me learn and grow as a lookout. Special thanks to Tim Klein, for the wisdom and

ACKNOWLEDGEMENTS

fatherly support, and for asking me all of the hard questions ("What will you do when you're alone with your mistakes?") My deepest respect for the legacies of lookouts who climbed long before me and those unique individuals who climb and look out with me today. You make me feel held and seen—it's the most unusual and yet beautiful kind of diaspora community. You all know who you are, but special shout-outs to: Justin, Caitlin, Christa, Tori, Dan, Aaron ("Sam"), Erin, Tamara, JD, Marina, and Shane. I've forever got y'all, five-by-five.

To Holly the Tower Dog, who waits beneath the tower, season after season, and yowls happily when I climb down. We speak a language of our own, but you deserve to be named here. Thank you for teaching me how to sink into the wild in a deeper, more animal way. How to calmly stand my ground in the path of wolves and bears and lynx. How to not fear Nature, but observe and respect her.

And finally, to the black spruce, muskeg, and boreal forest that's held me, tested me, and taught me. In an early draft, I wrote that "Nature is indiscriminate to love." After five short years learning from the land, I've realized that just isn't true. Nature needs love and shows love—if you're open, if you're listening. Gratitude to the non-human community, the plant life and animal life, that supports me in ways that are difficult to put language to but are felt skin and spirit deep. There has never been a more urgent time to listen to the voices of Indigenous peoples, who were here long before many of us, who steward these lands with a philosophy of reciprocity and regeneration.

I write these words from a place of solitude, but I am not alone— not really.

Gratitude to you all from the bottom of my heart.

Here's to another fire season together.

—XMA685, OUT.

READING LIST

Connors, Philip. *Fire Season: Field Notes from a Wilderness Lookout.* New York: Ecco, 2012.

Dillard, Annie. *Teaching a Stone to Talk: Expeditions and Encounters.* New York: Harper Perennial, 2013.

Ehrlich, Gretel. *The Solace of Open Spaces.* New York: Penguin Books, 1986.

Estés, Clarissa Pinkola. *Women Who Run With The Wolves: Myths and Stories of the Wild Woman Archetype.* New York: Ballantine Books, 1997.

Kimmerer, Robin Wall. *Braiding Sweetgrass: Indigenous Wisdom, Scientific Knowledge and the Teachings of Plants.* Minneapolis: Milkweed Editions, 2015.

Luke, Pearl. *Burning Ground.* Toronto: HarperCollins Canada, 2000.

Maclean, Norman, and Robert Redford. *A River Runs Through It and Other Stories.* Chicago: University of Chicago Press, 2017.

Meloy, Ellen. *The Anthropology of Turquoise: Meditations on Landscape, Art, and Spirit.* New York: Pantheon Books, 2002.

Neruda, Pablo, Ilan Stavans, and Firuz Kazemzadeh. "Solitude." Poem. In *The Poetry of Pablo Neruda.* New York: Farrar, Straus and Giroux, 2005.

Solnit, Rebecca. *A Field Guide to Getting Lost.* New York: Penguin Books, 2006.

Struzik, Edward. *Firestorm: How Wildfire Will Shape Our Future.* Washington: Island Press, 2019.

Tymstra, Cordy. *The Chinchaga Firestorm: When the Moon and Sun Turned Blue*. Edmonton: University of Alberta Press, 2015.

Forest Protection Branch. *The Lookoutman's Handbook*. Alberta Forest Service, 1971.

SOURCE NOTES

INTRODUCTION

7   *Despite the popular image*: Philip Connors, *Fire Season: Field Notes from a Wilderness Lookout* (New York: HarperCollins. 2011).

8   *For the past ten thousand years*: Cordy Tymstra, *The Chinchaga Firestorm: When the Moon and Sun Turned Blue* (Edmonton: University of Alberta Press, 2014).

8   *First Nations peoples in northern Alberta*: Amy Cardinal Christianson, "Burning Territory: Indigenous Fire Stewardship," Landscapes in Motion (website), March 5, 2019, landscapesinmotion.ca/updates-1/2019/2/26/burning-territory-indigenous-fire-stewardship#_edn1.

9   *In the northern boreal*: Mike Flannigan, Alan S. Cantin, William J. de Groot, et al., "Global Wildland Fire Season Severity in the 21st Century," *Forest Ecology and Management* 294 (April 15, 2013): 54–61.

9   *Scientists today agree*: Edward Struzik, *Firestorm: How Wildfire Will Shape Our Future* (Washington, DC: Island Press, 2017).

CHAPTER ONE

16   *After her disappearance*: Dawn Walton, "Police Still Seek Leads Five Years after Alberta Fire Spotter Vanished," *Globe and Mail*, August 25,

2011, theglobeandmail.com/news/national/police-still-seek-leads-five-
years-after-alberta-fire-spotter-vanished/article591970/.

17   *Indigenous women are 12 times*: John Paul Tasker, "Inquiry into missing
      and murdered Indigenous women issues final report with sweeping calls
      for change," CBC News (website), June 3, 2019, https://www.cbc.ca/
      news/politics/mmiwg-inquiry-deliver-final-report-justice-
      reforms-1.5158223/.

CHAPTER FOUR

59   *Studies show that, in human–bear encounters*: Stephen Herrero, *Bear
      Attacks: Their Causes and Avoidance* (Lanham, MD: Rowman &
      Littlefield, 2018).

CHAPTER SIX

92   *The RCMP believes*: Dawn Walton, "$700-Million Slave Lake Fire Caused
      by Arsonist," *Globe and Mail*, November 1, 2011, theglobeandmail.
      com/news/national/700-million-slave-lake-fire-caused-by-arsonist/
      article547260/.

93   *In 2002, a U.S. forest ranger*: Mindy Sink, "National Briefing: Rockies:
      Colorado: Added Term in Forest Fire," *New York Times*, March 6, 2003,
      nytimes.com/2003/03/06/us/national-briefing-rockies-colorado-added-
      term-in-forest-fire.html.

94   *Climate change has resulted*: B.M. Wotton, M.D. Flannigan, and
      G.A. Marshall, "Potential Climate Change Impacts on Fire Intensity
      and Key Wildfire Suppression Thresholds in Canada," *Environmental
      Research Letters*, 2017, https://iopscience.iop.org/article/10.1088/
      1748-9326/aa7e6e.

CHAPTER SEVEN

114   *Approximately ten people die*: Rachel Maclean, "Lightning the
      'Deadliest Summer Weather Threat' in Canada," CBC News (website),
      September 14, 2016, cbc.ca/news/canada/calgary/lightning-deaths-
      canada-thunderstorm-season-1.3754024.

CHAPTER ELEVEN

166   *Pine beetles burrow*: Anna C. Talucci and Meg A. Krawchuk, "Dead
      Forests Burning: The Influence of Beetle Outbreaks on Fire Severity and
      Legacy Structure in Sub-boreal Forests," *Ecosphere* 10, no. 5 (May 2019).

166   *Pine beetle outbreaks*: D.D.B. Perrakis, R.A. Lanoville, S.W. Taylor, and
      D. Hicks, "Modeling Wildfire Spread in Mountain Pine Beetle–Affected
      Forest Stands, British Columbia, Canada," *Fire Ecology* 10 (2014): 10–35.

166   *Historically, pine beetle populations*: Andrew Nikiforuk, *Empire of the
      Beetle: How Human Folly and a Tiny Bug Are Killing North America's
      Great Forests* (Vancouver: Greystone Books, 2011).

167   *One study in northern B.C.*: Gordon Hoekstra, "Fires Rip through B.C.'s
      Tinder-Dry Pine Beetle Killed Forests," *Vancouver Sun*, July 20, 2014,
      vancouversun.com/technology/Fires+through+tinder+pine+beetle+
      killed+forests/10047293/story.html.

CHAPTER TWELVE

180   *Stephanie landed a job:* Excerpts and paraphrasing from "Treeplanting"
      on *Vinyl Cafe Family Pack* by Stuart McLean (2011). Used by permission
      of the estate of Stuart McLean.

CHAPTER FOURTEEN

191   Melanophila acuminata, *a centimetre-long*: Cordy Tymstra, *The Chinchaga
      Firestorm: When the Moon and Sun Turned Blue* (Edmonton: University
      of Alberta Press, 2014).

194   *But according to a 2014 study*: D. Romps, J. Seeley, D. Vollaro, et al.,
      "Projected Increase in Lightning Strikes in the United States Due to
      Global Warming," *Science* 346, no. 6211 (November 14, 2014), 851–54,
      science.sciencemag.org/content/346/6211/851.

CHAPTER SIXTEEN

217   *1993, Maurizio Montalbini*: John Phillips, "Sociologist Emerges from
      Grotto," *Times*, December 10, 1993.

217   *Steven Callahan, a sailor*: Steven Callahan, *Adrift: 76 Days Lost at Sea*
      (Boston: Mariner Books, 2002).

CHAPTER NINETEEN

240   *On September 22, 1950*: Cordy Tymstra, The Chinchaga Firestorm:
      When the Sun and Moon Turned Blue (Edmonton: University of
      Alberta Press, 2014).

242   *Thanatologist Kriss Kevorkian*: Jordan Rosenfeld, "Facing Down 'Environ-
      mental Grief,'" Scientific American, July 21, 2016, scientificamerican.
      com/article/facing-down-environmental-grief/.

242   *Many scientists determined that*: Edward Struzik, Firestorm: How
       Wildfire Will Shape Our Future (Washington, DC: Island Press, 2017).

245   *Many wildfire scientists are advocating*: "Prescribed Fire," Alberta
       Agriculture and Forestry (website), April 27, 2017, wildfire.alberta.
       ca/prevention/prescribed-fire/default.aspx.

245   *Many scientists agree that we ought*: Thomas Fuller and Matthew
       Abbott, "Reducing Fire, and Cutting Carbon Emissions, the
       Aboriginal Way," New York Times, January 16, 2020, nytimes.
       com/2020/01/16/world/australia/aboriginal-fire-management.html.

EPILOGUE

290   *Cumulonimbus flammagenitus, or pyrocumulus*: Ed Struzik, "Pyro
       Storms: A New Danger in the Era of Wildfires," *Narwhal*, May 8, 2019,
       thenarwhal.ca/pyro-storms-a-new-danger-era-wildfires/.

TRINA MOYLES is a writer, photographer, potter, and seasonal smoke spotter in the northern boreal. She is the author of *Women Who Dig: Farming, Feminism, and the Fight to Feed the World*. Her award-winning writing has been published in *The Globe and Mail, The Walrus, Alberta Views, Maisonneuve, Hakai Magazine*, and many other publications. She lives, writes, and adventures in northwestern Alberta with her canine sidekick, Holly.